DATE DUE

DEMCO 38-296

U.S. Utility Mergers and the Restructuring of the New Global Power Industry

U.S. Utility Mergers and the Restructuring of the New Global Power Industry

Edward B. Flowers

QUORUM BOOKS
Westport, Connecticut • London

Library of Congress Cataloging-in-Publication Data

Flowers, Edward B., 1939–
U.S. utility mergers and the restructuring of the
new global power industry / by Edward B. Flowers.
p. cm.
Includes bibliographical references and index.
ISBN 1–56720–163–6 (alk. paper)
1. Electric utilities—Mergers—United States. 2. Gas industry—
Mergers—United States. 3. Electric utilities—United States—
Finance. 4. Gas industry—United States—Finance. 5. Investments,
American. 6. Investments, Foreign. 7. Electric utilities—
Mergers. 8. Gas industry—Mergers. 9. Electric utilities—Foreign
ownership. 10. Gas industry—Foreign ownership. I. Title.
HD9685.U5F59 1998
333.793′2′0973—dc21   97–46575

British Library Cataloguing in Publication Data is available.

Library of Congress Catalog Card Number: 97–46575
ISBN: 1–56720–163–6

First published in 1998

Quorum Books, 88 Post Road West, Westport, CT 06881
An imprint of Greenwood Publishing Group, Inc.

Printed in the United States of America

The paper used in this book complies with the
Permanent Paper Standard issued by the National
Information Standards Organization (Z39.48–1984).

10 9 8 7 6 5 4 3 2 1

Contents

Illustrations

Chapter 1

An Economic Environment That Encourages Mergers

This book describes close to $70 billion of global utility mergers stemming from the anticipated deregulation of the U.S. gas and electrical utilities industries. These mergers that occurred from 1995 to 1996 are completely restructuring U.S. power utilities. Thirty-seven billion dollars of these mergers, a full 53 percent, occurred abroad. About two-thirds of the foreign mergers were U.S. takeovers, while the remaining one-third was mergers, defensive and otherwise, of U.K. firms with other U.K. firms. Thus, this may be the first time a U.S. industrial restructuring has caused more investment abroad than domestically.

This book explores the diversity of strategies and changes driving these mergers and concludes that although complex, the mergers can be explained by strategies traditionally used in domestic mergers and acquisitions. These very large U.S. utilities now consider themselves to be operating in a global industry of private, deregulated utilities, and they are determined to survive through mergers that help them cut costs, spread expenses, and increase profits.

AN ENCOURAGING ECONOMIC ENVIRONMENT

"Consume thy rival" may be the new law of corporate survival in the U.S. utilities industry,[1] because the U.S. economy continues to provide an environment that encourages further mergers. In 1995 and 1996, the U.S. economy was characterized by slow, steady growth, with deep downsizing and an increase in global competition. In this environment, suppliers didn't raise their prices, and there were few new easy sources of profit. One strategy pursued by many corporations was to increase their profits by acquiring their competitors. Seventy-five percent of mergers from 1995 to 1996 were accomplished with stock swaps or stock swaps supplemented with cash. The soaring U.S. stock market put a premium on the value of corporate shares and made this type of acquisition cheap. As a result, mergers in all industries soared to $647 billion in value in 1996, three times the 1995 level. Although there may not be many large merger targets left, the economic scenario for continued mergers remains in force. Table 1.1 provides the recent history of merger activity in the United States.

The financial parameters of these utilities show that their shares, which under performed the S&P 500 over the past five years, returned less than half as much as average U.S. shares. Bonds issued by utilities are still well rated, with their senior debt at a median of A−, but the ratings are expected to drift down to BB or BB+ as the industry is deregulated. Still, no debt collapse is expected. Utilities might be considered to be doing well considering that they own $150 billion of worthless assets mostly in mothballed nuclear plants. These plants pose a vexing question: Who gets stuck with the bill for these white elephants? Regulators appear determined to help the utilities refinance these stranded costs as they gently ease the utilities into the world of competition.

THE CONTINUING MERGER BOOM IN ELECTRICAL UTILITIES

Three industries are expected to continue mergers unabated: commercial banks, electrical utilities, and industries merging on an international scale—pharmaceuticals and telecommunications.[2] Pharmaceuticals and telecommunications industries are rationalizing on a global scale, so they require more time to complete their combinations. The pharmaceuticals industry was a global oligopoly before the current merger wave began, whereas telecommunications needs to merge to provide their corporate clients seamless global communications coverage. Hospital combinations and the television broadcasting industry have largely exhausted the possibility for large mergers. Table 1.2 points to the industries

Table 1.1
Merger Activity in the United States

Year	Billions of Dollars
1988	$353.0[1]
...	...
1990	$182.6
1991	$138.4
1992	$150.1
1993	$234.3
1994	$341.9
1995	$518.8
1996	$658.8

Note: [1]Merger peak

Source: Securities Data Company, and *New York Times*.

Table 1.2
The Most Active Industries for Mergers

Industry	Billions of Dollars
Telecommunications	$103.89
Electric, gas and water distribution	**$ 41.06**
Radio and television broadcasting stations	$ 37.54
Business services	$ 32.00
Health services	$ 28.75

Source: Securities Data Company, and *New York Times*.

most active in mergers and acquisitions.

The data indicate that mergers in U.S. electrical and gas utilities mergers are second only to the mergers in telecommunications. In the electrical utilities industry, there have been so many mergers in gas and electricity companies that in December 1996 there was a backlog of nine mergers involving 19 companies, with $36.577 million in revenues, $88.57 billion in assets, and serving 11.3 million customers. All of these mergers need the approval of the Federal Energy Regulatory Commission (FERC). Responding to this backlog, FERC has recently announced that it intends to speed up its approval procedures by using a scoring procedure similar to that used by the Federal Telecommunications Commission (FTC).[3]

A unique feature of the wave of mergers in the U.S. gas and electrical utilities industries is that almost $70 billion of global utility mergers were caused by U.S. merger activity. The U.S. utilities mergers began in anticipation of the deregulation of the U.S. gas and electrical utilities industries by FERC and other regulatory boards. Table 1.3 shows the pattern of this spreading wave of global utilities mergers and acquisitions.

Forty-seven percent of the U.S. mergers, worth almost $33 billion, occurred between U.S. utilities while the remainder (53 percent) occurred in mergers with firms located outside of the United States. Included in the $37 billion of mergers with foreign power companies are $11.6 billion between British power utilities (not involving a U.S. partner). Some of these mergers were definitely defensive responses to U.S. takeover initiatives.

The private, deregulated utilities of the U.K. and Australia were the primary targets for U.S. utilities as they sought higher profits abroad. Whereas the U.S. utility industry has always been private, both Britain and Australia have recently received a double-shot incentive to greater profits—through privatization and deregulation. A total of $13.65 billion (19.5%) of U.S. merger money went to acquire British utilities, $4.97 billion (7.1%) went for Australian utilities, and $4.1 billion (5.9%) went for Latin American utilities and projects.

In the telecommunications industry international combinations can be explained by the need for seamless global communications coverage, but in the utilities industry, the reasons are more complicated. Utility mergers have involved

Table 1.3

Announcements of Mergers and Acquisitions in Utilities: 1992-1996

Acquirer → Acquisition	%	$ Billions	US→US	US→UK	Total US→ Abroad	UK→UK	Total Abroad
US → US	47.1	32.91	32.91				
US → UK	19.5	13.65		13.65	13.65		13.65
US → Australia	7.1	4.97			4.97		4.97
US → Latin Amer.	5.9	4.10			4.10		4.10
US → Asia	3.9	2.70			2.70		2.70
UK → UK	16.6	11.60				11.60	11.60
Total	100	69.92	32.91	13.65	25.42	11.60	37.02
Percent		100.00	47.10	19.50	36.30	16.60	52.90

Source: Compiled by the author from various issues of the *Wall Street Journal* and the *New York Times*.

a diversity of strategies to take advantage of cost spreading; economies of scale of management, economies of scale of production, and systems operations; changing technology in power marketing; and a diversity of techniques to take advantage of the increasing securitization of the gas and electricity commodities. These changes seem to be occurring simultaneously in a global utilities market. The emergence of this market in energy has been heralded by the spreading wave of U.S. utilities mergers which show the determination of the very large firms to survive through mergers that cut costs, increase profits, and pool resources.

NOTES

1. Charles V. Bagli, "Conditions are Right for a Takeover Frenzy," *New York Times*, January 2, 1997, p. C3; and Gene Marcial, "Everybody's Talking Takeover," *Business Week*, June 16, 1997, pp. 102-106.

2. *Ibid.*

3. Agis Salpukas, "U.S. Agency Moves to Ease Utility Mergers, with Deregulation and Price Competition, an Emphasis on Speed," *New York Times*, December 19, 1996, pp. D1, D5.

Chapter 2

The Deregulation of the U.S. Power Industry

GAINS FROM DEREGULATION

The primary inspiration for the current wave of mergers and acquisitions activity in the U.S. utility industry has been the determination of U.S. regulators to deregulate the production of electricity and gas. These utilities compose an industry that is larger than either the automobile or the telecommunications industry. Deregulation has forced the restructuring of this mammoth $210 billion industry which is composed of many state utilities which had previously enjoyed over 90 years of protection as regulated monopolies.[1]

Economic Stimulus: Cheaper Power

The deregulation of the utilities industry is expected to stimulate the U.S. economy by producing a 20 percent decrease in the nation's power bill. This will free up from $40 to $60 billion for more productive uses.[2] These expectations are based on the recent deregulation experiences of a number of other industries. Deregulation has, in each case, resulted in much lower prices. In the airlines industry, deregulation brought down prices by 30 percent. In the long-distance telephone industry, deregulation brought rate decreases of 50 percent, and in the trucking and freight industries, rates fell 30 to 50 percent. McKinsey & Co. has estimated that in 1996 there was a $150 billion oversupply of plants producing electricity in the United States.[3] With deregulation and the shakeout of inefficient producers of electricity and gas will come with it, it is reasonable to expect that the power industry will also produce at least the same magnitude of consumer savings that deregulation has produced in airlines, telephones, and trucking.

United Kingdom and Australia Lead

Other countries also have discovered these savings and are leading the U.S. in the worldwide trend to power deregulation. Argentina, Britain, Chile, New Zealand, Norway, and Australia have already deregulated their gas and electrical industries. This phenomenon is not limited to electricity and gas production. It is happening even in railroads. U.S. rail companies have been buying control of deregulated European and Latin American railroads. In these companies, the jobs were formerly thought of as workfare and the rail companies were not run as if it was important to make a profit. Now, under U.S. management, these railroads are run to make money, and are doing so by taking advantage of managerial innovations, such as computer routing, that will bring efficiency, lower costs, and greater profits.[4]

GOALS OF DEREGULATION

Lower Electricity Rates

The goal of deregulation is to get more competition and lower electricity prices. These savings can be impressive: as much as one-third of the total power costs for a business. As competition heats up, businesses, local governments, and power plants have been jockeying for positions of competitive energy cost advantage across the United States. Issues that used to be determined by the utilities commissions that regulated local utilities are now being determined in emerging national energy markets as municipalities shop around for good electricity rates. As these customers, formerly captive to state electrical monopolies, break free of their local utilities, disputes between the utilities and their customers are ending up in the courts, where judges and juries are being asked to determine whether old contracts are valid, and if valid, at what rates. In effect, the courts are being asked to decide who—company, stockholders, bondholders, taxpayers, or electricity customers—is obligated to pay for the huge investments the power companies have made in capital equipment.

Patterns of Deregulation

Some states are in a rush to deregulate, like Rhode Island and New York. Other states, like California and Pennsylvania, plan to deregulate gradually, targeting the year 2001 as the first year of full deregulation. The gradualist states want to give their utilities a chance to prepare for full market competition. Table 2.1 provides a summary of current deregulatory activity in seven states. The plans show that a number of states plan to raise rates in the short run and charge various other levies to finance the retirement of their utilities' obsolete, uncompetitive generating assets before the real competition begins.

Table 2.1
Individual Cases of Deregulation

California: A law allowing deregulation was passed in September. A special charge of about one-third of an average monthly bill will be levied to pay for $29 billion in power plant debt.

Connecticut: Three state reports have urged deregulation. To pay for existing power plants, a charge on monthly bills has been recommended for consumers who switch providers.

Massachusetts: Most of the state's utilities have agreed to accept competition by 1998. utilities will add a surcharge of 3 to 3.5 cents per kilowatt hour to pay for $12.5 billion in debt for existing plants.

New Jersey: The utility control board wants competition in the wholesale market. It has not addressed how to pay for power plant debt.

New York: The Public Service Commission wants to deregulate by 1998. Last October, it asked utilities to submit recommendations.

Pennsylvania: A new law allows deregulation to be phased in by 2001. Utilities can add a surcharge to customers' bills for existing power plant debt, but the amount will be decided case by case.

Rhode Island: The law passed in June 1996 deregulates utilities starting July 1997. Full deregulation will occur by July 1998. To pay for existing plant debt, a surcharge of 2.8 cents per kilowatt hour will be levied the first three years, dropping gradually to 0.5 cents after 12 years and ending in 25 years.

Source: Barnaby J. Feder, "The Nuclear Power Puzzle. Who Will Pay for a Generation of Expensive Plants?", *New York Times*, January 3, 1997, D1, D3.

UTILITIES VERSUS MUNICIPALITIES

As businesses and municipalities eagerly shop around for cheaper electricity, large utilities and power authorities have begun to restructure themselves and maneuver to become more competitive. In New York, the Niagara Mohawk Power Corporation has spent $4 billion buying out its expensive above-market power purchase contracts with small independent power producers.[5] On the other hand, the New York Power Authority (NYPA) will probably have to give up its subsidized power rates to upstate utilities, because in a deregulated market it would cause a conflict of interest—NYPA would be subsidizing its own competitors.[6] In Tennessee, the conflict is between a Virginia municipality that has, until recently, been locked into long-term, above-market electricity supply contracts with the giant Tennessee Valley Authority (TVA).

Bristol, Virginia Battles the TVA

Large Authorities, Small Municipalities

When the small municipalities of Bristol, Tennessee and Bristol, Virginia went shopping for cheaper power, they found themselves pitted against the Tennessee Valley Authority, one of the largest power authorities in the world. Although the two Bristols are twin cities, astride the Tennessee-Virginia state line, the Virginia town managed to fight its way free to cheaper electricity rates, while the Tennessee town could not. The struggle between these two small towns and the TVA illustrates a number of issues involved in the scramble for cheaper energy that is becoming available in decontrolled electricity markets.

These towns are two halves of a manufacturing town on the Virginia-Tennessee state line, which even share the same main street. Both towns bought their power from the TVA. But January 1998, Bristol, Virginia will cut its electricity costs by one-third by unplugging itself from their long-term power contract with TVA.[7] David Fletcher, chairman of the Bristol, Virginia Utilities Board, led the fight to cut its contractual ties with the TVA and contract for cheaper power. Fletcher grew up on a TVA "demonstration farm," in a severely economically depressed section of Virginia. But during the fight to get Bristol free to contract for cheaper power, he says he lost his naive nostalgia for the new deal institution of TVA.

Fletcher was promoting an industrial park to stimulate the local Bristol economy when he discovered that the planned park was close to the power lines of a competing electrical utility which offered electricity at rates 30 percent below those of TVA. But TVA refused to supply power at the lower rate and also refused to let the Bristol Utilities Board acquire a block of power from the other company at the lower price. Bristol felt that lower electricity rates were essential to make their industrial park competitive with other industrial sites in Virginia.

TVA later offered lower rates for new industry but only if Bristol would sign a long-term contract requiring them to give ten years notice of cancellation. The cancellation clause would renew itself automatically every year. With the unanimous support of the city's major industries, Bristol notified TVA that they intended to let their current four-year contract lapse and seek cheaper sources of electricity. TVA says it offered $7 million in benefits, but it held out for the ten-year cancellation requirement.

When faced with a similar challenge in Columbus, Mississippi, the TVA had refused to offer new industries the same discounts that they gave other distributors. Columbus sued, and TVA counterclaimed that if Columbus' four-County area left its customer base, Columbus owed them $65 million in order to pay for their stranded costs of underused equipment. TVA further threatened to move a $470 million coal-fired electricity plant employing 300 workers out of the Columbus co-op's area.

J. D. Bowie of Columbus's four-County Co-op, a lawyer, helped Fletcher plot Bristol's strategy. But in 1992, when Congress passed a law opening the use of power lines to competitive suppliers of power, TVA called on the 40-member Congressional caucus from its region to lobby for a TVA exemption from the law and got it. This exemption would have locked Bristol into its TVA contract, but Bowie drew up an amendment to the exemption, presented by Democratic Representative Rick Boucher of Virginia, which, although not mentioning Bristol by name, exempted utilities from the TVA exemption if they had electrical-service contracts that terminated before October 1991.

C. Farnham Jarrard, a member of the Bristol Utilities Board, then discovered that TVA had $27 billion in debt that was associated with half-built nuclear plants which would never be completed. When TVA advertised that it would supply power to new regions in Virginia at competitive prices, Jarrard reasoned that TVA would have to raise power prices on their old customers in Bristol, in order to pay the interest on the debt and supply power at cheaper prices to the new customers at the same time. This line of reasoning stiffened Bristol's resolve to find a new source of cheaper power.[8]   After surveying nineteen bids from other power companies, Bristol accepted a fixed price contract from Cinergy Corporation in Cincinnati which would deliver power over TVA's lines and cut the town's wholesale utility bill by $70 million over seven years. The total savings were double the city's annual budget. Bristol planned to use a third of this saving to lower local consumer electricity rates. The remainder of the savings would be invested in capital improvements, such as repairing leaking sewer lines. Then came the inevitable counterattack by TVA.[9]

By January, when Cinergy's selection was announced, Mr. [Craven] Crowell had already begun the TVA's counterattack. He held a meeting with the town's major industries there, telling them there were ways the TVA could connect with them directly. He wrote Jerry Wolfe, the Virginia Bristol's mayor, accusing Cinergy of "predatory" pricing, asserting that a new supplier could trigger "blackouts," and threatening a lawsuit for stranded costs that could wipe out all of Bristol's projected $70 million saving.

In March, TVA crews began surveying land near Bristol's industrial park for a new power line and a substation that could deliver TVA power directly. It offered some of the local industries—Bristol's biggest electricity customers—an even bigger bargain: Whatever price Bristol offers them, the TVA would undercut it by 2%.[10]

Although TVA attempted to lure away Bristol's industries with prices below Cinergy's—which TVA called predatory—Crowell said that this was "just good business." He explained: "They [Bristol, Virginia] can leave, but they can't take our territory with them."[11]

A Structured Conflict

The fight between Bristol, Virginia and TVA has also structured the broader conflict that is now developing between TVA and TVA's five biggest customers, the municipal power plants in Nashville, Chattanooga, Huntsville, Memphis, and Knoxville.[12] These municipalities account for 30 percent of TVA's total market, and have now banded together in a consortium calling themselves the Big Five.

The consortium is gearing up to get bids for cheaper power from outside suppliers just as Bristol did. Matthew Cordaro, president of the Nashville Electric Service, said that he expects Congress to try to carve up the giant TVA agency, suggesting that if TVA actually has low costs and will pass them on the consortium, the consortium will help TVA to survive. If they don't, they Cordaro thinks that there will be a wild scramble for cheaper power and everyone will get hurt.[13]

CHANGING ELECTRICITY PRICES CAUSE FINANCIAL RISK

Who Pays for Stranded Assets?

The price of electricity is often determined in a very dramatic way by deciding who pays for the stranded costs of an electrical utility's past investments in capital equipment. Under state regulation, most utilities were allowed to pass these costs on to their monopolized markets, but under deregulation, the utilities may have to eat some of these costs—to the dismay of their stockholders and bondholders. These issues are not trivial, because utilities are capital intensive operations, and their capital charges are large enough to cause bankruptcies if managed in the wrong way.

*New Rules May Bankrupt New
Hampshire's Northeast Utilities*

Northeast Utilities, a Massachusetts power utility holding company with $1.7 million customers in three states and the owner of New Hampshire's biggest utility, went to court on March 3, 1997, to block a state deregulation decision that would sharply limit the utility's practice of passing on costs of past investments to customers.[14] Northeast said that a February 28, 1997 ruling by the New Hampshire Public Utilities Commission (PNC) would slash the company's revenue by $341 million over two years and might ultimately drive it into bankruptcy.

Northeast attempted to stop that ruling, by seeking a temporary restraining order from the Federal District Court In Concord.[15] The stranded costs of New Hampshire's utilities were mostly associated with high-cost nuclear power facilities that had not proven capable of supplying competitively priced power. The ruling by the Public Utilities Commission was noteworthy, because it was the first time that a state regulatory agency had attempted to restrict a utility's right to recover their "stranded costs."

Steven Fetter, an analyst for Fitch Research, thought that the PUC was taking a harsh position by supporting the view that the utility could not get their customers to pay for the costs of their previous investments.[16] The legal move by Northeast to protect itself is likely to be a foretaste of conflicts to occur between utilities and their regulators as the pace of electrical power deregulation begins to pick up speed in the United States.

Who Pays for Poor Management ?

On August 8, 1997, the United Illuminating Company joined eight other utility companies in suing troubled Northeast Utilities. The companies claimed that they had suffered damages of more than $200 million that they said represented their costs involved in closing down the Millstone 3 nuclear power plant in Connecticut when the Nuclear Regulatory Commission (NRC) lifted its license.[17] The NRC has now closed all four of the Connecticut nuclear power plants run by Northeast Utilities.

Although United Illuminating had helped Northeast Utilities, it said that it had an obligation to their stockholders to recover the costs of closing the four plants. The shutdowns cost the companies over $200 million, which represented the cost of power purchased from other sources because Northeast could not produce the power with the closed Millstone 3 plant. United Illuminating said that it would continue to help Northeast get the four plants reopened.

When Is a Power Contract Enforceable?

Bankruptcy of an Electricity Giant

When two limited partnerships, the Okeelanta Power Company and Gator Generating Company, both filed for bankruptcy in the U.S. Bankruptcy Court for the Southern District of Florida, the filing threatened a default on $288 million of high-risk municipal bonds.[18] This was enough to roil the $1.3 trillion municipal bond market. The bonds had been sold to finance capital investments that allowed the power companies to produce electricity by burning the waste products from Florida sugar production.

Okeelanta and Gator filed for bankruptcy because of a suit by Florida Power & Light Company that sought to terminate an arrangement which would have required them to purchase power from Okeelanta and Gator under federal law.[19] Florida Power & Light said that it brought suit to protect its customers from paying the exorbitant electricity prices of Okeelanta and Gator.

Okeelanta and Gator claimed that they had satisfied the conditions of their power contract with Florida Power & Light. But Florida Power & Light claimed that the partnership's prices were exorbitant, that the plants' power supply was undependable, and said that the plants had missed a January 1, 1997, deadline to become fully operational.[20] Florida Power & Light also said that electricity purchased from these facilities was from 80 to 100 percent more expensive than power that they could obtain from other sources or by generating it themselves.[21]

If Okeelanta Power and Gator Generating both default on their bonds, the default could be as large as the 1983 default of the Washington Public Power Supply System; the mid-1970s default of New York City; and the 1994 petition for bankruptcy of Orange County, California, according to the Bond Investors Association data which has kept track of such defaults since 1980.[22]

Financial Risks

Municipal power companies have traditionally been regulated industries with bond coupon rates that were more or less guaranteed by state regulatory commissions. These commissions weighted the interests of residential consumers in getting reasonably priced electricity against the financial needs of the power companies. In the case of partnerships like Okeelanta/Gator, however, the interest payments on their project's bond financing was to be made from the revenue from long-term commercial power contracts. This arrangement is riskier than depending on the revenues from a large number of residential customers, because a small number of well-informed industrial customers, such as Florida Power & Light Co., often find their financial viability threatened when unanticipated decreases in the market price of electricity turn their long-term contracts into huge, long-term money losers. Whereas public utility commissions have stuck electricity consumers on Long Island with rates 400 percent above market for years, the payment of such costs by Florida Power & Light would simply result in insolvency and bankruptcy. This means that the capital financing of Okeelanta and Gator was very risky.

Because of the high risk in financing such projects, the bonds used to finance these projects may have low ratings or no ratings at all[23] and the bonds will carry very high coupon rates to compensate for these low ratings. The largest bondholders in the Okeelanta/Gator situation, however, were large municipal bond funds. These funds often purchase high-risk, high-return bonds like those of Okeelanta and Gator, in order to increase the return on the funds.[24] The bonds of the Okeelanta/Gator partnership were held by such firms as Dreyfus Co., Eaton Vance Management Corp., and Franklin Resources Inc. These firms braced themselves for what might be one of the largest bankruptcies in the municipal bond market history.

The situation with the Okeelanta/Gator bonds is symptomatic of the type of new financial risk in utility bonds that is emerging as the U.S. electrical utilities industry restructures itself. Utility bondholders generally support Chapter 11 bankruptcy filings because these filings tend to decrease the risk of total default on utility bonds. This is because the Chapter 11 proceedings allow the company to continue operating although it may be in financial default, opening up the possibility that cash flows from continued operations may be used to make the interest payments on bonds and mitigate what might otherwise be a total default under a Chapter 10 bankruptcy filing. The residential and commercial customers of Florida Power & Light, however, would prefer to pay the lower electricity rates available on the increasingly competitive national power grid. Individual cases such as the Okeelanta/Gator bankruptcy will obviously be decided on the facts that determine whether or not the long-term contract was breached or not.

The Local Impact of High-Cost Electricity

The case of the Village of Lynbrook, New York, located on the south shore of Long Island, affords an opportunity to see how a small municipality fared when it attempted to negotiate its independence from the high-cost electricity of the Long Island Lighting Company (LILCO). LILCO was known as the highest cost electricity producer in the United States. In the case of the Okeelanta/Gator bankruptcy, Florida Power & Light acted as a powerful intermediary agency as it attempted to avoid its high-cost, long-term power contract with Okeelanta/Gator in order to secure lower electricity prices for its customers. In the case of the village of Lynbrook, however, the elected officers of the Village had to pit their expertise against one of the largest power companies in New York state.

The Economic Background: The Long Island Economy Revives

Long Island boomed in the 1980s, and then crashed during the downsizings of the 1990s. But the LILCO built its Shoreham nuclear plant in the mid-1980s, when Long Island's economy was growing rapidly. The company thought it would be wise to invest in atomic power production to meet their forecast energy needs for the expanding Long Island economy. The state encouraged LILCO to build the Shoreham plant, and entered a contractual agreement with LILCO in 1989 allowing it to recover nearly all of its costs from the state. This contract remains as evidence of the state's part in building this notoriously costly plant.

Unfortunately, during the boom of the 1980s, the Long Island economy became very dependent on the military industry. The military component of Long Island's industry was typified by Grumman, the aircraft giant. When the cold war ended, and the military needs waned, the department of defense downsized the U.S. military-industrial complex. In 1992, Grumman left Long Island to merge with Lockheed, and the Long Island economy began to slide into recession. The area has recovered now, and the pace of economic growth on Long Island, is brisk again. In 1997, Long Island had one of the nation's fastest regional growth rates: there was only 3.5 percent unemployment, down from a high of 8.1 percent in 1992 when Grumman left.[25]

During this recovery, Long Island's industry began to moving in new directions. New growth was led by smaller companies that specialized in computer technology, retailing, and business services like law and accounting.[26] But the growth was uneven, and although jobs had been lost throughout the region, most of the recovery occurred only in Nassau and Suffolk counties. The new jobs, requiring high-tech skills, sprung up in Suffolk County's Hauppauge area, while in Hempstead, in Nassau County, buildings were still vacant.

Despite Long Island's new prosperity there remained fears that the local, economy would overheat and hit the wall like it did in 1988.[27] Local labor shortages had become a problem, and the mergers and buyouts among the computer companies were making employees nervous.[28] Although Wall Street bonuses bid up the high end of the Long Island housing market, the island was still too dependent on Wall Street, and the New York City economy was less broadly based than it was before the recession. The complex web of relationships that connect Long Island to the global economy made the area more competitive, but the new patterns of growth made it difficult to tell where people worked, either geographically or within the corporate hierarchy.[29] Labor statisticians found it difficult to locate the new jobs that had decreased Long Island's unemployment rate.

Lynbrook Goes Private

The Village of Lynbrook provides an example of Long Island's sense of unease with its newly enhanced global competitiveness. Lynbrook originally began investigating the feasibility of forming its own municipal electrical utility in July 1995, after mayor Eugene Scarpato received a call from the former mayor of Shoreham.[30] The Shoreham mayor asked Scarpato if he would be interested in retaining the services of a power consultant, Power Alternatives, Inc. After consulting with Power Alternatives, the board of Lynbrook voted unanimously to put the utility issue on their March 1997, referendum ballot. A favorable vote on the issue would have allowed the board to decide whether or not to form the municipal utility.

On February 5, 1997, Lynbrook held a town meeting where discussion based on the issues involved in establishing a local municipal electrical utility.[31] The meeting also explored the question of whether a Lynbrook utility could retail power to their contiguous communities (Greenport, Freeport and Rockville Centre) was also explored. The hearing was chaired by Lynbrook's mayor, Eugene Scarpato, who allowed LILCO, the chamber of commerce, Power Alternatives, Inc., and the Board of Trustees to present their cases.

LILCO's figures indicated that it would cost the village $157 million to create their own utility.[32] Power Alternatives, who did a feasibility study, said it would cost only $35 million. Lynbrook residents were paying about 17 cents per kilowatt hour for their electricity. Power Alternatives said that forming a municipal utility would lower that cost to about seven cents per kilowatt hour. At the February 5 meeting, Harry Levitt, president of the chamber of commerce, said that LILCO's high rates handicapped local businesses. Scarpato reported that the average bills for customers was $1,000 per year, but Steve Giordano, LILCO's representative, said the average LILCO bill for Lynbrook residents was only $961 per year.

Lynbrook Condemns Lilco's Lines

By Tuesday, March 18, 1997, Lynbrook residents were prepared to vote on a proposal to allow their village to condemn the transmission wires in a portion of LILCO's electric distribution system and assume the responsibility for providing their own electricity.[33]   An editorial in *Newsday* said that this would be a bad idea. *Newsday* argued that Lynbrook would pay more for electricity if it ran its own electrical utility, than it was currently paying through LILCO. The editorial gave these reasons:

- Lynbrook would have to condemn LILCO's power lines before it could use them, and would be required to pay replacement value for them, which was more expensive than their original cost, which was used as the basis for computing the rates LILCO charged Lynbrook.

- Since the amount to be paid for the lines would be decided in court, that amount was uncertain, and this uncertainty was like writing a blank check for the cost of the condemned equipment.
- Lynbrook would have to pay for its share of the power plants, power contracts, and other system facilities now serving the village, plus a portion of the costs associated with Shoreham. A recent ruling by the FERC explicitly required municipalities to reimburse utilities for such stranded costs.
- Millions more would have to be paid to separate Lynbrook's utility from LILCO and to set up their own power system.
- Lynbrook's self-service power operation could expect to have increasing costs at just the time that the Long Island Power Authority (LIPA) was expected to decrease Long Island electricity rates.
- Lynbrook had a relatively low demand for electricity, and virtually no commercial demand, so a Lynbrook utility would be a low volume, high-cost electricity provider, compared to the large, efficient volumes of power that LILCO deals in.
- Lynbrook would not have LILCO's expertise in power generating and transmission nor would the city be good at responding to emergencies, and this would make Lynbrook an undependable supplier.
- The village would forgo more than a million dollars in tax revenue that it received from LILCO each year.[34]

Put simply, the *Newsday* article argued that Lynbrook was neither large enough nor expert enough to be able to set up a low-cost power utility. And the equipment would be expensive, because LILCO would not be required to give away transmission lines—Lynbrook would have to pay for them. LILCO said it would be better to accept their tax rebate and any LIPA rate decreases that might come along.

The Village Point of View

At a Lynbrook Village meeting on March 3, 1997, Gary Vegliante and Joseph Prokop, the president and attorney of Power Alternatives, respectively, contradicted statements that had been made in a flier handed out by LILCO, point by point.[35]

- Since the FERC would consider the cost of stranded investments prior to condemnation, the village would know how much it would have to pay LILCO before going forward with condemnation of LILCO's wires and transmission facilities.
- Lynbrook would be not committed to a condemnation as LILCO said it would. If costs appeared to be too high, Lynbrook could walk away from the condemnation .
- Lynbrook could get firm power bids from electric companies throughout the United States. "Power is now available, but we are not entitled to get a bid without a license," Vegliante said.[36] Recent firm power bids had been below 3-cents per kilowatt hour.
- A dedicated work force, on-site in the village, similar to the one enjoyed by Rockville Centre, was to be included in Lynbrook's the plan. "You will not be waiting for

someone in a master office," Vegliante said. "Nationally, municipal utilities are much more reliable. Publicly owned utilities do a much better job."[37]

• Vegliante gave as an example the city of Messina, in upstate New York, which, after municipalizing its electric power over the past 10 years, had been able to electrify all-night ball fields for free. Messina had also continued to pay PILOT (payments in lieu of taxes) to its school board, and their electricity rates had continued to go down.

"As individual customers with LILCO," added Mark Choinacki, a Lynbrook resident, "Lynbrook residents have no bargaining power, but [with a municipal electrical utility] the village will get more leverage and better deals. It will be able to shop around. Lynbrook crews will be here faster. Common sense tells me I'm better off with group purchasing. Common sense tells me if we rely on LIL-CO, we can't touch them." Looking at the village board, he said, "I can get these [Lynbrook] people out of here in two years, but I can't touch LILCO."[38] Harry Levitt, president of the Lynbrook Chamber of Commerce said, "It's only the next step; to let LILCO win on this one is a slap in the face to the rate payers. Why shouldn't we have the right to at least explore. I'd love to see what FERC says."[39]

Jo-Anne Taormina, the LILCO representative who had waited patiently to speak until end of the meeting, suggested that the discussion was premature because it appeared that LIPA would take over a portion of LILCO, so that "you're going to realize a savings."[40]

The pro and con arguments in "opt out" negotiations of both Lynbrook and Bristol, Virginia show that the largest issues are whether or not and how much a locality would be required to pay for the sunk costs represented by the past capital investments of the utility they wished to leave. From the point of view of the utility, these investments were made in order to serve the locality that was opting out of their system. The utilities do not feel that localities that "opt out" should be able to strand the utility with unused capacity. But FERC has encouraged municipalities to shop around as a new national market for electrical power emerges.

In the case of Lynbrook, the largest capital charge that they might have to pay is their share of LILCO's enormously expensive Shoreham nuclear plant which was never allowed to go into service. It is easy to differ on the ways that the value of this stranded cost might be calculated. Moreover, LIPA's promise to reduce Long Island power rates was predicated on LIPA's plan to refinance the sunk costs of Shoreham in a way that would spread Shoreham's costs over 20 to 30 years. Lower annual costs for Shoreham would allow LILCO to reduce their electricity rates to a point where they might be able to compete with coal-fired market rates.

THE UNCERTAINTIES OF DEREGULATION

Deregulation will allow regulated energy prices to change. When these prices change, the new prices will throw into stark contrast which of the past capital investments of the utility companies were good, and which must be judged to have been bad. The financial risk of the stranded costs springing from the bad investments will fall not only on the utilities themselves, but also on their stockholders, bondholders, commercial customers, and on large groups of their residential consumers and the politicians that represent them.

Who Pays the Stranded Costs of the Shoreham Plant?

One of the most dramatic situations involving stranded costs was the LILCO's investment in the Shoreham nuclear plant. This plant saddled most of LILCO's Long Island customers with the high rates needed to pay the $20 billion in stranded costs for over eight years. The 800-megawatt Shoreham plant may be the most expensive thing ever built and then abandoned.[41] Thirty-two years ago, the plant was expected to cost $65 to $75 million; by the time it was completed in the mid-1980s and licensed in 1989, its cost had climbed to $6.5 billion. Shoreham began to drag down the Long Island economy as LILCO's escalating electric rates drove away thousands of jobs.

LILCO had intended to recover the costs of the plant over 40 years in its electricity billings, and had already charged its customers $5.6 billion for Shoreham costs and interest, but the plant never opened because of public concerns over safety. In 1989, Governor Mario Cuomo gave the plant to LIPA for a decent financial burial.[42]

The Main Issue: Future Costs

Paying the Shoreham costs forced LILCO's rates to be uncompetetitively high. The debt needed to finance the plant was expected to cost LILCO's customers $20 billion or more. Estimates of the future costs are shown in Table 2.2. Were LILCO to have been relieved of Shoreham's stranded costs, its electricity rates would have been only 18 percent of their current rates (17 cents per kilowatt hour).[43] This situation threatened LILCO's survival, because, in 1997, under the leadership of FERC, the national market for electricity was rapidly becoming a competitive market and businesses were locating where energy prices were cheapest. By 1997, the cost of Shoreham was billions of dollars in projected interest payments on the debt used to finance it. To clear this debt using electricity rates that were politically tolerable, meant repaying over a long period of

Table 2.2
Estimates of the Future Cost of Shoreham

Estimate	Background	$ Billions
Harry Davitian	President, Entek Energy Consulting Co.	$14.3
Gregory Palast	Economist directing Suffolk investigation of Shoreham	$25.0
n.a.	A critic's study	$20.0

Source: Bruce Lambert, "Hot Issue in LILCO Takeover Talks: Who Pays for Shoreham?," *New York Times* (March 16, 1996), pp. 41-42.

time. But interest also had to be paid on the bonded indebtedness necessary to finance the Shoreham debt over the tens of years during which the debt would gradually be paid down.

Various plans had been proposed to pay for Shoreham, but there were not many viable alternatives. The plant was not suitable for conversion to oil or gas. Even demolishing the plant would have been a major expense. Irving Like, proposed that Shoreham be made a monument, either to "man's technological stupidity, or as a memorial rest home for all of the politicians either elected or defeated because of it."[44] LIPA said that all it wanted was a fair settlement of the Shoreham costs that would reduce electricity rates.

The Long Island Power Authority

The LIPA was created to deal with the Shoreham problem. The authority was expected to issue long-term tax-exempt bonds that would refinance the Shoreham costs and spread these costs over a period that would allow LILCO to have a competitive price for its electricity.[45] The Brooklyn Union Gas merger with LILCO was also concocted to deal with this problem. LILCO had been wounded by Shoreham's costs and forced to enter merger talks with Brooklyn Union to gain access to cash flows sufficient to ensure its solvency and survival. Critics insisted that LILCO should absorb a large part of the Shoreham loss, but LILCO always opposed this demand.

Under the plan proposed by LIPA, the State of New York would allow the authority to reduce the size of payments on Shoreham's debt by refinancing the debt with tax-exempt bonds issued by the power authority.[46] This financing would require the largest such bond issue in the nation's history,[47] and the tax exemption feature of the bonds would allocate a portion of Shoreham's cost to every taxpayer in the state.

LILCO's Point of View

The Shoreham debt could also be shrunken by forcing LILCO to absorb a portion of the loss. This alternative had a strong equitable appeal for every Long Island resident or businessman who had suffered by paying LILCO's high electricity rates. The capital investment in Shoreham plant was, after all, a LILCO decision. But LILCO resisted this point of view.

LILCO contended that it had already paid enough by writing off $1.7 billion of Shoreham's costs in the 1980s, a penalty imposed by state regulators for the company's bad judgment in building the plant. LILCO pointed to their agreement with the state that they could recover all of Shoreham's costs, while critics pointed out that competitive changes under recent energy deregulation policies had prompted other utilities to accept losses on so-called stranded assets like the Shoreham plant.

LILCO further argued that when Shoreham was conceived, energy use was rising, Long Island was growing, and LILCO was facing brownouts and blackouts. Shoreham would have been a profitable investment but for a sequence of unforeseen events. First, there was the movement which stressed the environmental risk of nuclear plants, especially to those located on an island with no credible evacuation plan in case of a meltdown. Then the "China Syndrome" was dramatized in a disaster film. Then came an actual meltdown at Three Mile Island. This meltdown galvanized local opposition to the Shoreham plant, and prevented LILCO from ever putting the plant into service.[48] LILCO played their part in this disaster scenario by allowing the breakdown of their conventional diesel engines at Shoreham in 1983. Cynics nicknamed the diesel engines Snap, Crackle and Pop.

When the scenarios of conventional disaster had become mundane, financial disasters began to kick in. There were construction delays, cost overruns, double-digit inflation, and tougher regulations to contend with. Inflation pushed up construction costs to mammoth proportions. Joseph McDonnell, a LILCO senior vice president, said, "As you look back, mistakes were made by everyone it touched, both inside the company and in the political arena."[49]

What if Shoreham had been allowed to open? Charles Komanoff, an energy economist said, "Nuclear operating costs, exclusive of construction, are about the same as buying power from the grid, about 3 cents a kilowatt hour. Over all, it's a big fat wash."[50] But paying the current cost of the debt incurred by the Shoreham project made LILCO's price of electricity to Lynbrook (17 cents per kilowatt hour), more than 400 percent above market price.[51]

Who Pays for Shoreham?

In March 1996, the negotiations for a partial state takeover of LILCO, were reaching their final stages. Two main questions emerged: What was the true cost of Shoreham, and who should pay for Shoreham's stranded costs? The negotiations to decide who would pay for Shoreham's enormous expense involved a number of politically deceptive initiatives:

- The proposed merger of Brooklyn Union Gas Company and LILCO.
- The proposed refinancing of Shoreham by LIPA.
- The proposed state buyout LILCO, which included both LILCO's debts and assets—the stranded costs of Shoreham.[52]

Analysts were outspoken on the connection between these events: "This is all about the Shoreham debt—the merger is kind of a sideshow," said Ashok Gupta, an analyst for the National Resources Defense Council, an environmental group. "The underlying issue is do we have to pay for Shoreham or not? People are still fighting the old fight. It's like Bosnia, they have long memories."[53]

If the Shoreham mess was not enough to make New Yorkers cynical about their state's power policies, it was easy for them to be cynical about Con Edison's advertising slogan: "The company you know; the people you trust." This slogan rankled because New York residents knew they paid Con Edison 10.7 cents per kilowatt/hour, whereas electricity was selling for 6.9 cents nationwide. Moreover, the cost of electricity was far higher in New York's metropolitan areas.[54]

This did mean, though, that New Yorkers had a lot to gain from restructuring their electric industry, and New York politicians sensed this.

In an attempt to reduce these costs, on March 13, 1997, the staff of the state's Public Service Commission and Consolidated Edison agreed on a plan to restructure the giant electrical utility.[55] The plan would break Con Edison into three different companies that would all be operated by a single holding company. One company would generate power, another would retail power and the third would distribute power. The power distribution company also would be required to deliver the electricity of other, competing, generation companies, even if they were headquartered out-of-state. This plan was expected to set the pattern for the future development of all seven of New York's power utilities.

For the competitive features of New York's restructuring plan to work, its new power retailing companies had to be able to buy cheap power from out of state suppliers if it was available. But there is limited line capacity to carry cheap power into the state, and limited line capacity to carry it into New York City and Long Island. So about one-third of the power used in New York City and on Long Island would still have to be produced locally at relatively high-cost.[56] It remains an interesting question how much a court might fairly charge a condemnor for the power lines connecting New York City and Long Island with the rest of the world under the current conditions of scarcity.

New York state law also requires its utilities to pay for power from small gas-fired plants and small dams.[57] This power cost $1.4 billion per year more than power purchased on the open market. To end this situation, the Public Service Commission planned to buy out these small contracts at an estimated cost of $10.6 billion. Either way, the taxpayers or the electricity consumers, or both, would continue to pay the bill for a law that was inspired by the OPEC oil panic of the 1970s.

Mohawk Power Corporation and The New York Power Authority

In March 1997, Niagara Mohawk Power Corporation announced a buyout of $4 billion of the industry's $46 billion of above-market-price contracts with small independent power producers which had kept their prices uncompetitively high.[58] Without the agreement, Mohawk would have probably ended up in Chapter 11 bankruptcy.[59] With the agreement, the corporation would find itself in a stronger financial position, so the company asked the state Public Service

Table 2.3
Leader of the Pack

New York Power Authority	$40.49 Billion
Los Angeles Dept. of Water and Power	$22.07 Billion
Salt River Project (Arizona)	$21.95 Billion
South Carolina Public Service Authority	$16.02 Billion
Puerto Rico Electric Power Authority	$15.62 Billion

Source: American Public Power Association

Commission and its shareholders for permission to raise $3.5 billion to pay the cash part of the settlement. Although Mohawk suspended its current dividend, it intends to lower its electricity rates for the 1.5 million customers it serves upstate early in 1998.[60]

The New York Power Authority (NYPA) sold electricity to Niagara Mohawk Power Corporation as well as to three other upstate utilities. At the same time Mohawk was buying out its high-cost power suppliers, the NYPA found itself negotiating to sell its power for higher prices on the emerging national market.[61] NYPA is the largest publicly-owned state or local electric utility in the United States (see Table 2.3). NYPA also owns a third of the state's transmission system and operates some of the largest power plants in New York, including the huge Robert Moses Niagara hydroelectric power plant at Niagara Falls. The Robert Moses plant delivers about 10 percent of New York State's electricity and gives NYPA a distinct price advantage over New York's other electrical utilities.[62] This advantage is shown in Table 2.4.

Whereas Mohawk charged their average industrial customer 9.1 cents per kilowatt-hour, NYPA charged 4 cents or less to newer customers. But those who signed on in the past were paying just over a penny per kilowatt hour under statutory provisions mandating cheap power for certain industries.[63]

Because of its conflicts of interest, NYPA will probably be dismantled or reorganized as the national market in electricity becomes more competitive. This is partly because NYPA sells power to New York utilities at the same time that it competes with them. It is also because it is the state's lowest cost electricity producer, but is, at the same time, publicly financed. Harvey Shapiro, the president of the Energy Association of New York State, the trade and lobbying group for the state's private utilities, pointed out that, "You really can't have a tax-subsidized entity as a competitor."[64] Speaking in even broader terms, Gary J. Lavine, senior vice president of legal and corporate relations at Mohawk, agreed, "Sorting out the various roles that the authority has played as supplier, competitor, regulator and partner is a very complex and intricate process."[65] Because the au-

Table 2.4
A Price Advantage (Cents per Kilowatt Hour)

Lilco	17.7
New York Power Authority	**7.1**
Con Edison	10.4
New York Power Authority	**7.6**
New York State Electric and Gas	9.8
New York Power Authority	**6.1**
Niagara Mohawk Power	9.1
New York Power Authority	**5.3**
National Average	6.5

Source: New York Power Authority.

But by July, the warm reception for Pataki's plan had turned cold. The legislature felt that the $7.3 billion plan to take over LILCO and give Long Island customers rate rebates was too costly.[78] The administration also had been forced to give up large parts of its plan to cut back state welfare, including proposals to cut benefits to the poor and home care plans for the able-bodied unemployed.[79] Pataki hoped these measures would placate Republicans in the senate who were concerned that they might be deepening poverty and creating new social and administrative burdens for local governments.[80] In a two month period, the administration had gone from an attack into a broad retreat.

Critics began calling the plan to take over LILCO a subsidized windfall for the utility. The Brooklyn Union Gas Company, LILCO, and consumers groups all lobbied their positions intensely in Albany. "They said to me pretty straight out that there's a lot of wiggle room here—that was a surprise to me," assemblyman Robert K. Sweeney of Lindenhurst said of the lobbyists.[81] Previously, the proponents of Pataki's plan had indicated that the basic deal was inviolate. "It's a gold-plated bailout of LILCO shareholders at the expense of the rate payers," said Wayne Prospect, a consultant to the Hauppauge Industrial Association. He noted that LILCO's stock was up nearly 40 percent from its low point in 1996.[82]

Crucial Questions

Could the state back out of its 1989 agreement that allowed LILCO to recover nearly all of its costs from the Shoreham nuclear plant (by charging high rates)? The administration said no. But if the answer turned out be yes, the savings to consumers would be far greater.[83] Consumer advocates pushed for LILCO to swallow some of Shoreham's costs and lower its rates. This was equivalent to asking the stockholders to pay for Shoreham, and LILCO responded by telling LIPA and the Pataki administration that they were still bound by the 1989 agreement. "The State of New York has a contractual agreement to let LILCO recover the cost of Shoreham," said Louis R. Tomson, first deputy secretary to the governor and one of the architects of the buyout.[84] An attempt to break the pact would tie the state up in court for a decade, Tomson claimed. The state said the buyout would settle the matter once and for all, guaranteeing that Long Island ratepayers would pay for Shoreham.[85]

The deal was fairly sweet for LILCO and its shareholders. While this was galling to the utility's legion of critics, it appeared that LILCO would accept no less, leaving the state with little choice if it wanted to execute the takeover.[86] The agreement pumped up consumer savings in the early years by increasing the amount borrowed by more than $500 million and postponing some of the debt repayment for years. That increased the long-term cost of the deal—and the amount people would pay for electricity over the next three decades—by more than $1 billion. State officials made arguments for this arrangement, but critics said it was meant to bolster Pataki's short-term political fortunes.[87]

The settlement could have been paid off by adjusting electric rates over many years, at no overall increase in costs, but that idea was rejected by Suffolk County officials. So instead, LIPA proposed sending customers rebate checks and giving them large rate reductions in the first five years. To pay for that, the authority would borrow $462 million, to be repaid by Suffolk residents through higher rates until the year 2028.[88] "The whole thing is just nutty," said Larry Shapiro, staff attorney for the New York Public Interest Research Group, a consumer advocacy group that opposes the takeover. "You borrow half a billion just to hand people some cash?"[89]

Critics charged that LILCO was being vastly overpaid for its assets. For instance, the LILCO's Nine Mile Point 2 plant, one of the most trouble-plagued nuclear plants in the country, had always been a money loser. Yet the LIPA proposed to pay LILCO $301 million, book value, for LILCO's share of the plant.[90]

Critics also claimed that the part of the agreement that had LIPA pay the merged LILCO-Brooklyn Union for the right to buy its power for 15 years hindered competition. By making the merged company uncompetitive in the long run. Around the country, monopolies were being replaced by competitive electric utilities with lower rates, and eventually LILCO-Brooklyn Union would have to compete with them rather than sell to LIPA.

Evidence of competition was not hard to find. On July 15, 1997, Calenergy made a $1.92 billion hostile bid to take over the New York Electric & Gas Corporation.[91] If the takeover was successful, it would give Calenergy 804,000 customers in upstate New York and provide the company a northeastern base to serve the surrounding states of Pennsylvania and Massachusetts they begin to open their markets to competition.[92]

Still, it appeared that the deal would, at least over the first decade, save ratepayers money. LIPA could issue tax-exempt bonds, so it would pay lower interest rates on the debts that LILCO has been paying. And unlike LILCO, the authority would not have to pay federal income tax on shareholder dividends. "All the people who complain that we're paying too much don't take into account that we're getting out of a $1.5 billion obligation for $625 million," boasted Tomson.[93]

In July 1997, the deal still needed the approval of FERC, and a waiver by the Internal Revenue Service (IRS) of $2 billion in capital gains tax that LILCO would otherwise owe.[94]

Takeover Details

Under the takeover plan, LIPA would buy almost all of LILCO's outstanding stock for more than $2.8 billion, and would take over $3.2 billion in LILCO debt. To pay for this and related expenses, the authority would sell $7.3 billion in bonds to be repaid by hydroelectric ratepayers over the next 33 years. The authority would acquire the company's transmission and distribution system, as well as LILCO's 18 percent ownership of the Nine Mile Point 2 plant. The state calculated that LILCO would be paid book value for its assets, or $2.5 billion.[95]

The merger agreement with Brooklyn Union provided that if the state took over LILCO's debt, LILCO stockholders would get a 9.6 percent bonus—more than $2 a share—in the form of a larger share of the merged company. But even with the promise of this bonus, LILCO's stock had under-performed the market as a whole since late 1996.[96] Although the price of LILCO's stock was, in effect, still in negotiation as the plan was being announced,[97] most of their stock-holders were upbeat about the proposed buyout because LIPA's anticipated offer for the stock had driven its price from $16 up to $24.[98]

One of the more appealing aspects of the deal was that it allowed Suffolk County, the town of Brookhaven, and the Shoreham-Wading River School District to escape $1.1 billion in judgments that LILCO had recently won against them for years of excessive property taxes on the Shoreham plant. Under the takeover, LIPA would inherit LILCO's role, agree to settle the Shoreham case for $625 million, and drop the other cases.[99] LILCO had several similar suits pending against other local governments that were potentially worth another $400 million.[100]

Takeover Approved

On July 27, 1997 the $7.3 billion deal for a partial state takeover of the LIL-CO was approved by the PACB. The deal involved the largest municipal bond sale in history.[101] The takeover effectively ended the lengthy dispute that had centered on Shoreham.

The PACB approved the state's takeover of all of LILCO's debts and many of its assets. The debts included the cost of building Shoreham and the assets included the Shoreham nuclear power plant itself. The merged LILCO and Brooklyn Union Gas Company would keep the other power plants, although the merged company would have an option to buy them in a few years. Under the agreement, LIPA, a state agency, would also acquire LILCO's transmission and distribution system.

The deal gave the merged company a guaranteed income by hiring it to operate the power system for at least eight years, and by paying for it, for years, for the option to buy the power it produces, though not necessarily for the power itself. While the agreement promises a 14 percent rate cut, the Pataki administration predicts average rate reductions of 19 percent over the first 10 years of the deal. Critics questioned whether even a 10 percent cut is attainable.

"This is nothing more than a bailout of LILCO by the state and the rate-payers of Long Island," said Gordian Raack, executive director of the Citizen's Advisory Panel to LILCO. "The rate savings guarantee," he said "is unenforceable."[102] To help soften this kind of resentment, Democrat Sheldon Silver obtained a written guarantee from the Pataki administration that the buyout would lower electric rates for Long Island by at least 14 percent over 10 years. Silver stated that although the guarantee was not legally binding on LILCO, ratepayers would be able to use the guarantee to sue LILCO if LILCO did not live up to its provisions. The guarantee would also have considerable political force.[103]

Brooklyn Union also promised to invest $1.3 billion—two-thirds of the cash the deal would leave the combined company with—on Long Island. Much of that money would go into developing connections for natural gas that could compete as a power source with electricity.[104]

On the August 7, 1997, the last annual meetings of the Long Island Lighting Company and the Brooklyn Union Gas Company, as separate companies were held. The stockholders of both utilities overwhelmingly ratified a merger of the two companies that was expected to be completed in 1998.[105]

The meeting represented an approval of one part of Pataki's plan for a partial state takeover of LILCO. The merger created a major new utility with revenues exceeding $4 billion a year, although the name and headquarters of the merged company have yet to be chosen.[106] The company's service territory would stretch from Staten Island through Brooklyn, most of Queens and nearly all of Nassau and Suffolk Counties—an area with more than six million residents and thousands of business customers.[107]

NOTES

1. Peter Coy and Gary McWilliams, "Electricity: The Power Shift Ahead," *Business Week*, December 2, 1996, 78-82.

2. Robert J. Samuelson, "The Joy of Deregulation; A Bit Here, a Bit There and Soon We're Talking Big Savings, Say, $50 Billion a Year," *Newsweek*, February 3, 1997, 39; and reports on the studies of Robert Crandall (Brookings Institution) and Jerry Ellig (George Mason Univ.), 1997.

3. Coy and McWilliams, "Electricity: The Power Shift Ahead," 78-82.

4. Anna Wilde Mathews and Jonathan Friedland, "U.S. Railroads are Making Tracks Overseas," *New York Times*, May 21, 1997, p. 4.

5. Gordon Fairclough, "Niagara Mohawk to Pay 19 Producers $4 Billion to Alter, End Energy Pacts," *Wall Street Journal*, March 11, 1997, pp. A1, A3; and Agis Salpukas, "Utility Seeks to End Costly Pacts With Power Suppliers; A move to Shed Contracts that Have Raised Power Rates," *New York Times*, March 11, 1997, pp. D1, B8.

6. Agis Salpukas, "When Electricity Goes Private, Deregulation May Change New York Power Authority," *New York Times*, July 11, 1997, pp. D1, D2, D3.

7. John J. Fialka, "Using Savvy Tactics, Bristol, Va., Unplugs from a Federal Utility," *New York Times*, May 27, 1997, pp. A1-9.

8. *Ibid.*

9. *Ibid.*

10. *Ibid.*

11. *Ibid.*

12. *Ibid.*

13. *Ibid.*

14. Agis Salpukas, "Northeast Utilities Sues to Block Move by New Hampshire," *New York Times*, March 4, 1997, pp. D8.

15. *Ibid.*

16. *Ibid.*

17. Bruce Lambert, "Utility Sued Over Millstone 3 Closing," *New York Times*, August 8, 1997, pp. B15.

18. Charles Gasparino, "Electric Giants File for Bankruptcy Protection," *New York Times*, May 19, 1997, pp. A1, A3.

19. *Ibid.*

20. *Ibid.*

21. *Ibid.*

22. *Ibid.*

23. *Ibid.*

24. *Ibid.*

25. Kirk A. Johnson, "Revived Long Island Finds Life after Military Contracts," *New York Times*, July 7, 1997, A1, B6.

26. *Ibid.*

27. *Ibid.*

28. *Ibid.*

29. *Ibid.*

30. Elizabeth Cooper, "Powerful Questions; BOT, LILCO Answer Questions, Comments at Public Hearing," *Lynbrook Local News*, February 6, 1997, pp. 1, 27.

31. *Ibid.*

32. *Ibid.*

33. Jo-Anne Taormina, "Letter from Lynbrook Village Community Relations; Vote on a Proposal To Allow the Village To Condemn a Portion of Lilco's Transmission Wires," pp. 1-2.

34. "Nassau County: A Bad Deal for Lynbrook," *Newsday*, March 13, 1997.

35. Andrea S. Halbfinger, "Study in Contrast, Lynbrook Village Meeting has Music, Cute Kids and Conflict," *Lynbrook Local News*, March 6, 1997, pp. 1, 3, 26.

36. *Ibid.*

37. *Ibid.*

38. *Ibid.*

39. *Ibid.*

40. *Ibid.*

41. Bruce Lambert, "Hot Issue in Lilco Takeover Talks; Who Pays for Shoreham?" *New York Times*, March 16, 1996, pp. 41-42.

42. *Ibid.*

43. *Ibid.*

44. *Ibid.*

45. *Ibid.*

46. *Ibid.*

47. *Ibid.*

48. Richard Perez-Pena, "Lilco's Hard Journey, Road to a State Takeover Began with Debacle of the Shoreham Plant," *New York Times*, July 21, 1997, p. B4.

49. Lambert, "Hot Issue in Lilco Takeover Talks," pp. 41-42.

50. *Ibid.*

51. Cooper, "Powerful Questions," pp. 1, 27.

52. *Ibid.*

53. *Ibid.*

54. *Ibid.*

55. *Ibid.*

56. *Ibid.*

57. Richard Perez-Pena, "Rate Cut Questions," *New York Times*, March 16, 1996, p. 42.

58. Jeff Bailey, "Niagara Mohawk Plan Is a Small Step Toward Easing Utilities' Power Woes," *Wall Street Journal*, March 12, 1997, pp. A1, A4, and Salpukas, "Utility Seeks to End Costly Pacts," pp. D1, B8.

59. Agis Salpukas, "Niagara Deal with Independents Could Reduce Price of Electricity," *New York Times*, July 11, 1997, pp. D1, B5.

60. *Ibid.*

61. Salpukas, "When Electricity Goes Private, pp. D1, D2, D3.

62. *Ibid.*

63. *Ibid.*

64. *Ibid.*

65. *Ibid.*

66. *Ibid.*

67. *Ibid.*

68. *Ibid.*

69. *Ibid.*

70. *Ibid.*

71. Richard Perez-Pena, "In Plan to Cut L.I. Utility Rates, Biggest Savings Come Early", *New York Times*, March 27 ,1997; and "Plan for Cut in Power Rates," *New York Times*, May 20, 1997, pp. A1, D1.

72. Bruce Lambert, "Stockholders of Lilco and Brooklyn Union Ratify Merger of Companies," *New York Times*, August 8, 1997, pp. B15.

73. Perez-Pena, "In Plan to Cut L.I. Utility Rates."

74. *Ibid.*, "Plan for Cut in Power Rates," pp. A1, D1.

75. *Ibid.*

76. *Ibid.*

77. Richard Perez-Pena, "Is this the End of Lilco? Silver Has the Crucial Vote," *New York Times*, July 13, 1997, pp. D1, B1, B7.

78. Bruce Lambert, "Warm Reception Turns Cold for Pataki's Plan to Take Over Lilco and Cut Rates," *New York Times*, July 17, 1997, p. B4.

79. Raymond Hernandez, "Pataki is Willing to Give Up Parts of Welfare Plan," *New York Times*, July 3, 1997, pp. A1, B4.

80. *Ibid.*

81. *Ibid.*

82. *Ibid.*

83. Perez-Pena, "Is This the End of Lilco?" pp. D1, B1, B7.

84. *Ibid.*

85. *Ibid.*

86. *Ibid.*

87. *Ibid.*

88. *Ibid.*

89. *Ibid.*

90. *Ibid.*

91. James P. Miller and Steven Lipin, "CalEnergy Launches Another Hostile Bid," *New York Times*, July 16, 1997, pp. A1, A3, A4.

92. Agis Salpukas, "$1.9 Billion Hostile Bid for Utility," *New York Times*, July 16, 1997, pp. D1, D18.

93. Perez-Pena, "Is this the End of Lilco?" pp. D1, B1, B7.

94. *Ibid.*

95. *Ibid.*

96. *Ibid.*

97. *Ibid.*

98. Lambert, "Stockholders of Lilco and Brooklyn Union Ratify Merger," p. B15.

99. Perez-Pena, "Is This the End of Lilco?" pp. D1, B1, B7.

100. *Ibid.*

101. Richard Perez-Pena, "State Officials Approve Partial Takeover of Lilco," *New York Times*, July 27, 1997, pp. D1, B6.

102. *Ibid.*

103. *Ibid.*

104. *Ibid.*

105. Lambert, "Stockholders of Lilco and Brooklyn Union Ratify Merger," p. B15.

106. *Ibid.*

107. *Ibid.*

Chapter 3

The Strategies and Causes of U.S. Utilities Mergers

DOMESTIC MERGERS TO INCREASE THE RATE OF RETURN

Acquisitions for Consolidation

As U.S. utilities merge to acquire market share and increase their rate of return the most basic pattern of merger is that of acquisitions to consolidate a utility's power specialty. One firm will sell off a power utility that is not a part of its core business, while another, generally larger, utility will acquire that unit in order to consolidate its market position in a power specialty.

When NGC Corporation acquired the power utilities of Destec Energy in February 1997, NGC made these acquisitions to consolidate its position as a major power producer.[1] Destec was 80 percent owned by Dow Chemical Corporation, which had previously sold 20 percent of Destec in order to raise their own share price. The final sale of Destec to NGC was done specifically to allow Dow to focus more on their core strategies in chemicals and thereby raise the value of their corporation. NGC was a large, national natural gas wholesaler with extensive operations in Texas and California, and for them, the purchase of Destec's power operations acted to consolidate their core competencies in energy.

Another variety of consolidation to achieve energy competency is illustrated by CalEnergy's attempt to take over the New York State Electric & Gas Corporation (NYSEG) to use it as a marketing platform to extend their energy sales. On July 15, 1997, CalEnergy announced its $1.9 billion hostile bid to acquire NYSEG, a utility that serves 804,000 customers in upstate New York.[2] NYSEG sells both electricity and natural gas in western, central, and eastern New York State, with service areas that include Binghamton, Elmira, Auburn, Geneva, Ithaca, and Lockport, where electricity rates are much higher than in other parts of the country.[3]

The combination is unusual because CalEnergy, based in Oklahoma, was used to taking risks in an unregulated environment, whereas New York State Electric was a local regulated utility concentrating on its share of state energy business. The easiest explanation for this odd combination in this relatively low priced deal is that NYSEG is expected to give CalEnergy a platform from which to launch energy sales not only in New York, but also in its adjacent states, Pennsylvania and Massachusetts, and the entire northeast region.

The CalEnergy-NYSEG deal is the latest in a series of utility mergers that breaks the old pattern of mergers between utilities with common geographical borders. The recently completed merger of Enron-Portland General (in July 1997) and the recently initiated merger of PacifiCorp-Energy Group—as well as the CalEnergy-New York Electric bid—show that the geographical distance between utilities is becoming less important, because energy resources are becoming more mobile. These new mergers have generally been friendly mergers done in order to strengthen the companies as they prepare for more deregulated competition.

CalEnergy: Competition, Rate Cuts
and International Acquisitions

In December 1996, CalEnergy began its recent series of acquisitions by acquiring Northern Electric PLC for $1.7 billion. Northern was a deregulated British energy utility which distributed electricity to 1.5 million customers.[4] Northern was an attractive acquisition target because Britain had recently deregulated its energy utilities, which in turn increased the potential profitability of all U.K. utilities. This merger showed a confident and internationally aggressive CalEnergy pursuing a high return foreign investment strategy similar to other U.S. national power marketing utilities.

By 1997, CalEnergy had concluded its British acquisitions and was interested once again in U.S. mergers. In July 1997, CalEnergy's bid for Northern was followed by its bid for NYSEG. CalEnergy's first approach was through informal negotiations with NYSEG's managers, seeking a friendly merger. But these overtures were followed the next week by a $1.9 billion hostile bid for controlling interest and an offer to assume $1.5 billion of NYSEG's debt in return for the remaining 90 percent of NYSEG's stock. David L. Sokol, the chairman and chief executive of CalEnergy, said that if his company won control of NYSEG, it would quickly begin to bring down the electricity and gas rates for New York customers in NYSEG's service area.[5]

NYSEG was viewed as a sluggish performer.[6] The New York Public Service Commission had let NYSEG increase its rates 2.9 percent in 1995 to offset their rising production costs. But another reason for allowing this increase was that NYSEG's profits were being hampered by New York's legal requirement that they purchase electricity from New York's high cost independent power suppliers.[7] Many New York utilities had begun buying out these high-priced contracts.[8] NYSEG had not done this, but instead filed a plan with the commission to reduce the independent power rates by 3 percent before deregulation went into effect in 1999.[9]

Sokol promised that under CalEnergy management, NYSEG would provide larger rate cuts and do it much faster than it otherwise could. The rate cuts would be paid for with funds from cuts in dividends, advances in materials technology, and the sale of more services—with no layoffs. Sokol also said NYSEG would provide CalEnergy with a marketing platform to sell electricity and natural gas throughout the United States as states opened up their utilities to competition.[10]

CalEnergy's core operations were once geothermal facilities in Southern California. The California operations tap underground steam to generate electricity, which is sold to local utilities. But from 1995 to 1996, the company had diversified its energy production interests by acquiring, first, Magma Power Co., its main competitor, in a $957 million takeover in 1995, and then in 1996, Northern Electric PLC, a major British electricity generator. The acquisition of Northern Electric was expected to provide CalEnergy with "a skill base we can use back in the U.S., in Europe, and Asia."[11]

Two-Stage Hostile Takeover

On July 15, 1997, CalEnergy announced an offer of $24.50 a share in cash ($159 million) to purchase up to 9.9 percent of NYSEG,[12] the most that they could acquire without federal and state regulatory approval.[13] CalEnergy said that it was willing to pay $27.50 ($1.9 billion) and assume $1.5 billion of NYSEG's debt in return for the remaining 90 percent of NYSEG in a friendly merger. NYSEG's shares jumped 15 percent to $3.1875, a 52-week high of $24.5625. NYSEG said that its directors would meet to review the bid after it began.[14]

The takeover offer came a week after NYSEG had brushed off CalEnergy's undisclosed buyout offer. Sokol had met with Wesley W. von Schack, the chairman, president, and chief executive of NYSEG to discuss a possible merger. Von Schack responded that the company's board had decided that a merger discussion was not a priority. Sokol did not interpret the reply as being an outright rejection, but he did find it confusing, and started the July 15 tender offer.[15]

It was clear that CalEnergy's ultimate goal was to acquire all of NYSEG's common shares.[16] Although CalEnergy's preference was for a friendly settlement, if NYSEG refused to negotiate, CalEnergy said it would ask NYSEG's stockholders to oust the current board through a direct solicitation, or at the annual meeting in the spring.[17]

CalEnergy's offer for 9.9 percent of NYSEG's stock was a partial tender offer. This type of takeover bid is especially well adapted to hostile takeovers because it keeps the pressure on the target company's board without triggering "poison pill" defenses or requiring a response to state regulations restricting the outright purchase of more than 10 percent of a target's shares. The purchase of the entire company would have to be approved by state officials as well as the Federal Energy Regulatory Commission.[18]

In this case, however, the partial tender offer backfired. Without regulatory approval CalEnergy couldn't buy more than 9.9 percent of NYSEG's shares. Unfortunately, it had offered only $24.50 for these shares, whereas it had promised to pay $27.50 for the remaining 91.1 percent of the shares.[19] Investors didn't want to tender their share for the lower price, and when CalEnergy received a poor response, they threatened to pull out of the deal, leaving those who had purchased at the lower price with a potential loss if the shares should fall below $24.50. CalEnergy would not disclose by what margin their bid had fallen short.[20]

"[CalEnergy has] found a company that is the right size for them that is right smack in the middle of this region," said Barry Abramson, a utility analyst for Paine Webber, but, he added, CalEnergy did not need to buy a utility, and appeared to be merely pursuing a good deal.[21] CalEnergy's first offer of $1.9 billion was close to NYSEG's book value, whereas utilities generally bring offers of one and a half times book value.[22]

In August 1997, CalEnergy abandoned their bid for NYSEG, saying that "the only beneficiaries in this process appear to be NYSEG's board, investment bankers, and lawyers, who will split over $20 million in fees."[23] NYSEG had not sought a "white knight," because NYSEG did not have the growth potential to satisfy that kind of merger,[24] but instead had "engaged in a scorched earth campaign which, in the span of just a few weeks, wasted an enormous amount of money that could have been used to create value for shareholders and lower rates for customers," according to the federal court judge who had overseen the hearings where NYSEG unsuccessfully attempted to restrain CalEnergy's offer.[25]

Sokol had effectively presented CalEnergy as "a catalyst for change," and, added J.P. Morgan Securities analyst Kyle Rudden, "CalEnergy is smart, definitely aggressive, and definitely well versed not only in M&A but in hostile acquisitions."[26]

But Wall Street had always viewed CalEnergy's offer too low,[27] although they thought that the offer would be difficult to turn down because NYSEG has been a sluggish performer, "[T]he numbers being thrown out there are significant premiums" over NYSEG's recent stock price, said Rudden.[28] "I don't think they lose anything because the offer of $27.50 appeared to be a very low offer compared to the company's asset value," said Barry Abramson, a utility analyst for Paine Webber.[29] CalEnergy "came in with a lowball price and didn't get it," said Paine-Webber analyst Bert Kramer.[30] By moving on to other options CalEnergy "shows they have a disciplined approach to making acquisitions," said Michael Worms, a utility industry analyst for First Boston.[31]

As the smoke cleared, NYSEG reached an agreement with the New York Public Service Commission to lower their energy rates, and, because of this, their stock values did not fall to their pre-offer levels.[32] CalEnergy said that its price cuts would have come sooner and would have been more than three times the amount of that agreement.[33]

PUBLIC-PRIVATE JOINT VENTURES:
WHEN PUBLIC ELECTRICITY GOES PRIVATE

Even before any large mergers had occurred in New York, the state had the New York State Power Authority (NYSPA)—a ready-made utility combination.[34] This Authority, a public-private joint venture, currently acts as supplier, competitor, regulator, and partner, and has been called to help the state during energy emergencies. In the mid-1970s it also helped to shore up the finances of Con Edison by acquiring and completing two plants.

In some ways, NYSPA could be a superb competitor. The authority owns 35 percent of the state's high voltage transmission lines and three important generating plants: The Franklin D. Roosevelt plant, a hydroelectric marvel near Niagara Falls, and two nuclear plants, the James A. Fitzpatrick, on Lake Ontario near Oswego, and Indian Point 3, in northern Westchester County. The huge Roosevelt plant, alone, supplies 10 percent of New York's electricity. What is more important is that these plants generate electricity efficiently. NYSPA's electricity rates are from 23 to 42 percent cheaper than electricity produced by New York's seven other private power utilities.

But in other ways NYSPA is merely a welfare agency masquerading as a utility. NYSPA sells electricity at *below* market rates to New York generators who are required by state law to buy electricity at *above* market rates from independent power producers.[35] NYSPA also sells cheap electricity to companies like Praxair, which manufactures industrial gases, and Cascades Niagara Falls, Inc., which makes cardboard from recycled materials. These companies operate in economically depressed areas, under conditions that would leave them insolvent except for the subsidy provided by NYSPA's unrealistically low electric rates. NYSPA charges only 4 cents per kilowatt hour to new customers qualifying for state aid, and older aid customers get rates as low as 1 cent per kilowatt hour. These rates compare favorably with the 17 cents per kilowatt hour paid by Long Island residents to the Long Island Lighting Company. Although this sounds like corporate welfare, even some New York Republicans feel that NYSPA should not be dismantled because of its value in attracting and retaining jobs for the state.

Under policies of market deregulation, however, New York is expected to become a fully competitive electricity market by 1999. "The way to make this [market competition] take place is largely by dismantling the power authority," said Larry Shapiro, the senior lawyer for the New York Public Interest Research Group.[36] Many deregulators are not sure that the authority—or any version of it —can remain a force under deregulation.

NYSPA's operations have glaring conflicts that would become more apparent in a deregulated market. The authority competes with three upstate utilities that are its customers. These utilities resell the authority's power without a profit to some rural customers. Activities like these would only distort electricity rates in a deregulated market. In addition to these problems, NYSPA has special advantages over private companies: it gets low cost state financing and it is exempted from taxes. In terms of marketing, the seven independent power companies have been restricted, until now, to their separate parts of the state, but with deregulation, they may sell power wherever they wish, both in and out of state, and other energy marketers are expected to come pouring in from out of state.

It is now certain that NYSPA's role will change and various scenarios have been suggested for the agency. These proposals have implications for other states with huge federal power systems, systems like the Tennessee Valley Authority and the Bonneville Power Administration in Oregon.

NYSPA might take over the other 65 percent of the state's transmission system and regulate fair and open access to the transmission grid. To avoid conflicts of interest, it would have to spin off its generating plants to guarantee neutrality. The flaw in this plan is that the purpose of deregulation is to reduce electricity rates, and it doesn't make good sense for the state to spin off, to Enron or the Southern Company, two large out-of-state energy producers, the low cost plants that NYSPA already owns—they already produce electricity cheaply. A better solution might be to create a separate public agency to keep the Roosevelt plant in the state's hands. Besides being a valuable generating asset, it is considered valuable to New York's economic development.

But others, like Shapiro, are not sure that the authority, or any version of it, can remain a force under deregulation, and it is unclear how the authority is to participate in what is supposed to be a fully deregulated, competitive energy market.

INTERNATIONAL ACQUISITONS

A 1988 study by John Doukas supports the idea that foreign utility takeovers are an effective way to pursue higher profits. Doukas looked at the effect of corporate multinationalism on shareholders' wealth using evidence from international acquisitions.[37] As far back as 1988, he found that the announcement of international acquisitions resulted in significant positive abnormal returns for shareholders of multinational corporations (MNCs) not already operating in the target firm's country. Unlike the current situation, Doukas found insignificant positive abnormal returns were experienced by the shareholders of American firms that expanded internationally for the first time, while insignificant negative abnormal returns were experienced by the shareholders of MNCs already operating in the target firm's home country. If the U.S. firm had expanded into a new industry and a new geographic market—especially markets less developed than the U.S. economy—the abnormal returns were larger. Shareholders of MNCs found the greatest benefits to be from foreign acquisitions where there was a simultaneous diversification across industry and geography.

Doukas's study predicts mixed results for U.S. utilities investing in the U.K. or Australia for the first time. The study, however, was published in 1988, whereas most of the mergers and international utility strategies explored in this chapter occurred in the most recent wave of utility mergers from 1995 to 1997. The details of this latest wave of mergers tends to show that U.S. utilities acquired foreign utilities either (a) to gain efficiencies in operating in the emerging markets for international power, (b) to get higher rates of return associated with faster growing foreign energy markets, or (c) to escape domestic merger restrictions intended to restrict national industrial concentration. Although many of these utilities invested abroad for the first time during this latest wave of mergers, most of the companies had considerable experience with large domestic mergers before moving abroad, and moreover, the global spread of utilities is occurring so quickly, with so many international acquisitions, that these companies can no longer be considered inexperienced multinationals.

The pattern of utility mergers directed by FERC in the United States allowed the mergers of many small local utilities, and also allowed the mergers of some very large non-contiguous utilities that had formerly produced dissimilar products: electricity and gas. These large power marketing mergers led to the formation of a top tier of firms that aggressively entered various state markets on a national scale to compete with state monopolies in gas and electric production as soon as the local utilities were deregulated. Sometimes a power marketing firm would acquire a local utility or a decontrolled local monopoly to use as a state or regional marketing platform. What followed was: the spread of U.S. utilities mergers abroad in search of greater return on investment.

Mergers Increase
Return on Investment

After $41 billion of mergers, U.S. utilities found that large merger partners were increasingly scarce in the United States. They also found that rates of return were higher in overseas power industries that had already been deregulated. Although the U.S. utilities industry is privately owned, it has not been deregulated, so from the point of view of regulation, foreign utilities were more advanced than those in the United States—and more profitable. The electricity and gas utilities in both the U.K. and in Australia had been recently privatized and deregulated, and rates of return were higher in these countries. The higher rates available abroad spurred U.S. foreign direct investment in gas and electrical power firms, especially in the U.K., Australia, and in Asia.

The most popular countries for U.S. utilities' acquisitions were the U.K. and Australia. Most of the U.K. takeovers were done by American firms: the Southern Company (Atlanta), Central Southwest Corp. (Dallas), Entergy, General Public Utilities (NJ), Cinergy (Cincinnati), CalEnergy, and Dominion Resources Inc. (see Table 3.1).

Although the British government had already decided to deregulate and privatize their utilities, the U.K. takeovers were often hostile. In the U.K., the Department of Trade and Industry, the Monopolies and Mergers Commission, and the Takeover Panel presided over mergers and acquisitions. Under Britain's process of gradual privatization, the government's "golden shares" in a British utility gave it an effective veto over any merger or takeover.[38] But utilities became fair targets for takeovers and mergers when the Department of Trade and Industry began to give up its golden shares under the government's policy of privatization.

Britain's Monopolies and Mergers Commission continued its push for privatization in the face of a wave of hostile U.S. takeovers despite acknowledging that the mergers of two giants in their privatized electricity industry might operate against the public interest.[39] Nevertheless, the government recommended that both the bid by PowerGen for Midlands Electricity and the bid by National Power for Southern Electric, both British-with-British mergers, be alowed to proceed.

U.S. takeovers of British utilities reached a peak late in 1996. On December 18, 1996, Entergy, a New Orleans-based utility company, succeeded in taking over London Electricity for $2.1 billion.[40] On December 19, British regulators cleared a £1.6 billion bid by Dominion Resources Inc. of Richmond, Virginia for East Midlands Electricity PLC,[41] and the previous week saw the approval of a £923 million offer by CalEnergy Inc. (Nebraska) for Northern Electric PLC, in a bitterly contested hostile takeover.[42] The CalEnergy takeover of Northern followed that of seven other regional electricity suppliers that lost their independence following the privatization of the electricity supply industry in 1990. Three of the companies were bought by American firms.[43]

Although the CalEnergy bid for Northern Electric, PLC, was ultimately successful, it was bitterly fought by Northern. CalEnergy's bid[44] was successful only after the decision by Britain's Takeover Panel to extend the takeover period so that 50 percent of the stock could be purchased by CalEnergy.[45] This decision came after the Takeover Panel learned that Barclays de Zoete Wedd, an investment bank that was advising Northern Electric, had been offered a £250,000 performance fee in conjunction with its business on behalf of Northern. CalEnergy asserted that the fee amounted to a bribe to get the investment bank to buy Northern Electric shares on the open market and block the hostile bid. The Prudential Corporation had bought more than 700,000 shares on the open market to block the bid of CalEnergy, and offered to match the £6.50 per share CalEnergy bid, saying that CalEnergy's bid was low. But CalEnergy soon increased its ownership from 39.77 to 50 percent of Northern's shares to complete its successful bid for Northern. CalEnergy stock rose $.75 to $30.00 on the New York Stock Exchange.

The Southern Company Takes
Over an Asian Power Producer

Whereas Britain and Australia attracted investments because of the value to be unlocked by the recent deregulation of these markets, other foreign countries, like Hong Kong, had higher rates of return because of faster rates of economic growth and greater demand,[46] relative to supply, than the oversupplied U.S. power market. "It offers us an opportunity to capture growth in the world's second-largest integrated utility market," said A. W Dahlherg, chairman of the Southern Company about the acquisition of Consolidated Electric.[47] British utilities also found Asian power markets attractive. When British regulators prevented the merger of National Power and PowerGen, both invested heavily in Asia, Australia, Europe, and the United States.[48]

Table 3.1

Recent U.S. Bids for British Power Companies

Date	U.S. Company	British Power Company	Amount
Jul 1995	Southern Company (Atlanta)[1]	South Western Electricity PLC (U.K.)	$1.65B purchased
Nov 1995	Central Southwest Corp. (Dallas)[2]	Seaboard PLC	$2.53B
Nov 1995	Entergy[3]	Citipower Ltd.	NA
Apr 1996	Southern Company (Atlanta)[3]	Southern Electric PLC (U.K.)	NA bid $10B+ bid
Apr 1996	Southern Co. (Atlanta)[5]	National Power PLC	dropped May 1996
May 1996	General Public Utilities (NJ) and Cinergy (Cincinnati)[6]	Midlands Electricity PLC	$2.B bid
Oct 1996	CalEnergy[7]	Northern Electric PLC	$1.55B
Nov 1996	Dominion Resources Inc.[8]	East Midlands Electricity PLC	$2.15B
Jun 1997	PacifiCorp[9]	Energy Group, PLC	$6B

Notes: [1] Matthew C. Quinn, "Southern Co. Pleased with British Venture in Natural Gas," *Atlanta Constitution*, April 30, 1996, p. D2; "British Utility Rejects Offer by Southern," *New York Times*, July 12, 1995, pp. C3, D3; Emory Thomas, Jr., "Southern Company Bids for Firm Is Rebuffed," *Wall Street Journal*, July 13, 1995, pp. A3, C22; "Southern in Hostile Bid for British Utility," *New York Times*, July 13, 1995, pp. C16, D16; "South Western Electricity PLC," *Wall Street Journal*, July 28, 1995, pp. B2, A6; "British Utility Fights Bid by Southern Company," *New York Times*, July 29, 1995, pp. 19, 37; Lawrence Ingrassia, "Southern Company to Pursue

Table 3.1 (continued)

Its Hostile Offer for British Utility as Another Bid Flops," *Wall Street Journal*, August 13, 1995, pp. A7, A6; and Richard W. Stevenson, "British Utility to be Acquired by Southern Company" *New York Times*, August 26, 1995, pp. 17, 33.

2 Nicholas Bray and Dawn Blalock, "CSW Bids $2.53 Billion for U.K. Utility; Dallas Power Company Cites Easier Rules in Britain, Tighter Market At Home," *Wall Street Journal*, November 7, 1995, p. A19.

3 Dawn Blalock, "Entergy Intends to Buy CitiPower Australian, Electric Utility," *Wall Street Journal*, November 20, 1995, p. A5.

4 "Southern Company Expansion," *Wall Street Journal*, November 2, 1993, pp. A11, A13.

5 "Southern Seeking British Power Concern," *New York Times*, April 10, 1996, D11; Southern Co.: Plan to Purchase Utility Dropped after U.K. Action," *Wall Street Journal*, May 9, 1996, p. B3; Matthew C. Quinn, "British Utility Seeks Own Merger in an Effort to Follow Southern Co.," *Atlanta Constitution*, April 13, 1996, p. B2; Matthew C. Quinn, "Southern Co. Stalks Major Acquisition," *Atlanta Constitution*, April 17, 1996, p. B1; Matthew C. Quinn. "British Power Producer Spurns Southern's Overture," *Atlanta Constitution*, April 19, 1996, p. C1; and Matthew Quinn, "Southern Co. Drops Attempt to Buy Second British Utility," *Atlanta Constitution*, May 9, 1996, p. E7.

6 Benjamin Holden, "GPU, Cinergy May Acquire British Utility," *Wall Street Journal*, May 7, 1996, p. A3; and "A U.S. Bid Seen Today for Midlands of Britain," *New York Times*, May 7, 1996, p. D7.

7 Agis Salpukas, "CalEnergy Offers to Buy British Utility," *New York Times*, October, 19, 1996, pp. D1, D7; "British Utility Lifts Payout in Face of Bid," *New York Times*, December 11, 1996, p. D6; and "CalEnergy Bid For Utility Is Victorious. British Distributor is Latest Acquisition," *New York Times*, December 25, 1996, pp. D1, D3.

8 "Dominion Agrees to Buy Power Utility," *New York Times*, November 13, 1996, pp. D1, D3.

9 Agis Salpukas, "PacifiCorp is Said to Reach Deal to Buy British Utility," *New York Times*, June 12, 1997, pp. D1, D7.

The Southern Company's October 1996 bid to take over Consolidated Electric Power Asia Ltd. (Hong Kong) is an example of an international merger that was made purely to increase the rate of return for the parent company. "The (Southern) company was betting that Asia's strong economic growth would continue to increase their demand for electricity," analysts said.[49] The investment in Consolidated was felt necessary to boost Southern's rate of return above the average for the electrical power industry over the next few years.

Southern had already exhausted its prospects for mergers in the U.K. and sought a rapidly growing power company with greater profit potential than that available through a U.S. merger partner. Consolidated was one of Asia's largest independent power producers. Consolidated supplied China and the Philippines, and had projects under development in Pakistan and Indonesia. The Southern Company was the parent of Alabama Power, Georgia Power Company, Mississippi Power, and the Savannah Electric Company. Despite the geographically separate markets, this merger appealed to the financial dictates of Wall Street.

Although Southern acquired Consolidated to enhance their growth, in 1995, Asian growth paused. Rapid Asian capital investment was often wasteful and some said that their "hard work, saving and scant consumption couldn't last."[50] But now, research at the Brookings Institution indicates that productivity gains in Asia have increased more than was previously thought.[51] This sets the stage for continued Asian growth in power, because, unlike productivity gains from capital investment, growth derived from increased productivity does not come at the expense of current consumption. The Southern Company has probably gambled correctly on their Asian power investments.

HUNTING MONOPOLIES IN A DEREGULATED WORLD

U.S. Deregulation and
Low Rates of Return

As U.S. utilities gird themselves to sell electricity at market-determined rather than commission-determined rates they have been merging to form more competitive marketing and production units. Part of the reason for these mergers is to become large enough and profitable enough that Wall Street would be willing to finance their growth and competition in world markets. At least one large U.S. utility, the Southern Company, has looked abroad for expansion when regulators denied them U.S. merger partners. The Southern Company was simply looking for higher rates of return to satisfy its stockholders and investment bankers. But in both the U.S. and in Great Britain, when utilities looked abroad for merger partners, they found better deals than they had at home.

The net result was the transformation of the U.S. utilities industry into a national market with fewer, larger, more competitive global utility firms, rather than a traditional utilities industry populated by many regulated state monopolies. These larger companies were concentrated in an upper tier of companies known as the national power marketing utilities. These Wall Street-financed firms were much more conscious of their rates of return than were the smaller utilities and attempted to keep these rates high by actively merging with synergistic, efficient utilities both in the U.S. and abroad.

Although the U.S. utilities industry was regulated for over 90 years, its rates of return have been historically low. The rates have been so low that over significant periods of time the industry's average rates of return have been lower than the industry's cost of capital. In a 1991 study, David Kaserman found that for U.S. electric utilities, the realized rates of return were less than the permitted rates from 1977 to 1982 and less than the market cost of capital from 1979 to 1982.[52] Permitted rates were also less than the cost of capital in 1980 and 1981. This means that utilities have to seek above average rates of return in foreign mergers to compensate for their lower than average rates of return on domestic operations.

Estimates of the effects of rate-of-return regulation on the behavior of electric utilities have assumed that the allowed and realized rates are equal in equilibrium and that the allowed rate is at least as great as the cost of capital. The validity of these assumptions was tested by examining rates of return and costs of capital over the period from 1973 to 1982, which found that the allowed rate of return *was* greater than the realized rate after 1976. This lays the blame for substandard rates of return squarely on the shoulders of U.S. utility executives. It was almost certainly the utilities' decisions to invest in a disastrous amount of high cost nuclear generation that pulled down utility rates of return in the United States so dependably in the postwar years.

Regulated Markets
with High Electricity Rates

Although Germany favors open markets, its electricity rates are among the highest in the world. German cartel authorities have been trying to make their energy markets more competitive to reduce these prices,[53] but these high rates recently attracted a takeover offer for the cartel that supplies Berlin with electricity. In fact, the reason that German utilities often have high rates is because their utilities are often monopolies. The consortium that made this bid included the Southern Company.[54] If this takeover is the forerunner of other American investments in Germany utilities it seems that U.S. utilities might be looking for monopolies in Germany that they can no longer find in the deregulated U.S. market.

On May 23, 1997, the Southern Co. formed a consortium with two German utilities that appeared ready to bid $1.7 billion for a controlling interest in the electrical utility that served the city of Berlin, Berliner Kraft & Licht AG, known as Bewag.[55] That would mark Southern as the first American company to crack Germany's monopoly utility market. To do this, Southern would take a $759 million position in the consortium for a 23.6 percent interest in the Berlin utility.[56]

The Atlanta-based Southern Company had been one of the first U.S. utilities to look abroad for acquisitions. This strategy was different from that of most other U.S. utilities. Most U.S. utilities facing deregulation merged with other nearby utilities in order to cut costs and get more control over a larger customer base. But since Southern already had a huge customer base and low cost power, it was in a good position to sell its power in other areas—as well as trading it in other countries—as soon as utility monopolies were stripped away in the United States. It was thought ironic then when Southern bought into Berlin's utility cartel. "It is a major problem," said an official of the German Federal Cartel Office, which feels that the merger would give the consortium a near-monopoly in the Berlin area. Ulrich Hartmann, chairman and chief executive of Veba, said the consortium would face a thorough German antitrust review because Veba already owned two other utilities in the Berlin area.[57]

Southern may have thought that they needed a cartel's pricing power to give them more control over the profitability of their German utility, because although Germany may intend to deregulate, Michael Sayers has estimated that a 33 percent workforce reduction is still necessary for Europe's utilities to just equal the average productivity of the U.K.'s electric sector after only five years of privatization.[58] Southern was already the leader of the U.S. utilities that had expanded abroad. Now that it has expanded aggressively into the British, Asian, and South American utility markets, the German bid indicates that they probably intend to increase their European presence in advance of the planned deregulation there.[59]

STRATEGIC POSITIONING WITH ENERGY ARBITRAGE

In many of the latest domestic and international mergers, the purpose of the merger twines together the desire for greater profit and the determination to acquire new marketing and trading skills. These elements are so tightly bound that it is hard to say whether fear or greed is responsible for the acquisitions. If the rates of return of the utilities fall, markets will allocate them inadequate amounts of capital needed for the next phase of utility marketing. This next phase of competition will be marketing-based competition occurring in deregulated markets.

Even after having spent over $20 million on brand name advertising, UtiliCorp-Peco was unable to establish for themselves a nationally recognized trademark. In addition the advertising, they found it necessary to form strategic alliances with AT&T and ADT corporations to get the job done.[60] The experience of UtiliCorp-Peco makes plain that national utility marketing in the United States is a very expensive proposition, and this means that larger utilities will have the financial economies of scale to compete better.

Another reason for international mergers is the feeling of utilities managers that if their firm is not able to trade the international energy commodities markets—to arbitrage coal, gas, and electricity contracts internationally—that they will lose so much operational efficiency that they will become uncompetitive.

Pacificorp's Bid for Energy Group

In June 1997, PacifiCorp reached an oral agreement to acquire Energy Group PLC of Britain for $5.8 billion in cash.[61] This was a high price to pay for Energy Group, but the price may be worth it if the acquisition gives PacifiCorp the right strategic position. Based in Portland Oregon, PacifiCorp is a major provider of electricity well beyond the seven Western states where it has 1.3 million customers.

PacifiCorp uses coal-burning plants to generate low-cost electricity and plans to market this electricity nationwide in states open to market competition.[62] Energy Group's extensive coal assets would fit in with PacifiCorp's strategy of buying cheap coal to produce low-cost electricity, but PacifiCorp's acquisition was based on a broader strategy than cost cutting.

PacifiCorp intends to take its strategy beyond buying raw materials cheap and selling power expensively by developing an ability to sell their excess electricity, gas, and coal in high-cost power areas such as the East Coast.[63] PacifiCorp wants to buy coal wherever it can get the best price, and it wants to sell its electricity wherever in the world it can get the best price. To do this more smoothly, PacifiCorp developed contracts for the various commodities they deal in—coal, gas, and electricity—that they can swap with their customers in different markets and different countries. Their acquisition of Energy Group signaled their intent to extend these deals overseas.

Energy Swapping Skills

PacifiCorp's bid to acquire Energy Group is an example of a large utility that wants to be able to swap contracts in any of the commodities they deal in—coal, gas, and electricity—across both markets and countries.[64] So far only large utility companies can swap these long term contracts because the contracts are not standardized for trading.

Although the power contracts are not standard, and the energy commodities have not yet been securitized, they may soon be. FERC would like to see a national market in U.S. electricity that would enable power to be purchased at a national price off the grid. To securitize these commodities for efficient trading, standard futures contracts would need to be developed for electricity. However, contracts for electricity are not now standard. Indeed, although Enron is competing in New York against Con Edison by providing cheaper electricity rates, Enron's commercial customers still find it difficult to compare the terms of Enron deals with offers made by Con Edison. The contracts are complicated. Despite these developmental problems, Enron has already developed a trading division that can efficiently handle complex electricity trades.[65] But Enron, intends to stay clear of PacifiCorp's niches in coal-based marketing.

Energy Arbitrage

PacifiCorp is pressing to extend their ability to arbitrage their own contracts for coal, gas, and electricity globally. Their acquisition of Energy Group is evidence of this. In one example of this freewheeling, market approach, PacifiCorp could swap coal contracts for contracts in electricity. PacifiCorp's Peabody Coal unit now has about 150 contracts to supply American power plants with coal.[66] PacifiCorp could renegotiate these contracts so that these plants would provide PacifiCorp with their cheap excess electricity production, which PacifiCorp could then sell in the Eastern United States at a premium.[67] In other words, PacifiCorp would rather sell electricity to its electricity-generating coal customers if it can make more money doing this than it can by selling coal to these electricity generators. This is how PacifiCorp is now arbitraging their corporate contracts for coal, gas, and electricity despite the absence of a global market in these commodities. A low-cost electricity contract is traded for a high-valued coal contract.

PacifiCorp's energy contracts could just as easily be traded by units in the U.S. and Britain, as within units all located in the United States. Britain is a good place for PacifiCorp to expand: it is more fully deregulated than the U.S., British coal is cheap, and the British energy market is open to foreign investment. Pacifi-Corp's purchase of Energy Group also signals that they would like to arbitrage energy contracts internationally from the U.K. market.

The international arbitrage of energy contracts is a new phenomenon being brought about by the current globalization of the world's energy utilities. This story is parallel to the developmental pattern of other world industries since World War II. The big story of the world's industries in the last decade, however, was the emergence of global business and markets from the activities of groups of multinational corporations. Once industries had global scope, capital markets began to emerge in most nations. As national stock markets developed into global capital markets, the large energy-trading utilities operating in many national energy markets began to knit these markets into a single global energy market. Thus, the multinational energy utilities' international arbitrage of energy contracts is one activity serving to globalize the world's energy industry.

The acquisition of Energy Group's extensive operations—the world's largest coal operations in private hands—gave PacifiCorp control of 26 mines in the United States that were part of the Group's Peabody Coal unit. These resources will allow PacifiCorp to better adjust its coal shipments and power transmissions to demand in various markets as U.S. deregulation of the energy market progresses. These coal operations would allow PacifiCorp to sell inexpensive Eastern coal to high cost Midwestern generators. The cheap coal could either be delivered locally (if Midwestern firms generate electricity locally) or PacifiCorp could burn the coal in its own generating plants to produce cheap electricity to be put on the grid for the account of a Midwestern generator. The use of this "coal-by-wire" type of energy arbitrage is spreading to other utilities, not only because it earns nearly risk free profits, but it also saves on rail transportation costs. "You decide what's the optimal thing to do—move the coal or move the energy," says Joseph T. Pokalsky, senior vice president at Southern Co.'s trading and marketing unit.[68]

Energy companies with large coal holdings also have a practice called "coal tolling" which provides a traditional background for the type of arbitrage developing out of coal's convergence with electricity. In a coal-tolling deal, a power-marketing energy company will purchase coal and ship it to an unaffiliated generating company which will convert the coal into electricity that the power-marketer then sells under its own brand name. Louisville Gas & Electric Corporation (LG&E) and PacifiCorp are pioneers in the coal-tolling area. Fifty-six percent of all U.S. electricity is generated by coal-fired plants, and much of the coal firing these generating plants comes from the low-sulfur coal in the Powder River Basin where PacifiCorp owns mines.[69] The acquisition of Energy Group added more Powder River Basin coal assets to PacifiCorp's assets. Vitol, SA of London is an example of a British power utility which has a Boston-based power marketing unit which engages in coal-tolling.[70]

The increasing arbitrage of energy has created a phenomenon known as "coal-energy convergence," "gas energy convergence," or simply "energy convergence." PacifiCorp CEO Frederick W. Buckman said that the most successful energy companies, "will come to see gas as 'liquid electricity' and coal as 'solid electricity.'"[71] New energy trading techniques make trading, arbitrage, and hedging of coal, gas, and electricity possible. Recent changes in the federal laws designed to open up wholesale power markets have aided this type of trading.[72]

Although energy trading is not yet securitized, it does constitute an arbitrage process that allows hedging. As the deregulation of markets proceeds, and companies continue to make long-term energy contracts under conditions of fluctuating electricity prices "one has to find ways of hedging risk," says J. Craig Baker, vice president of power marketing and trading with American Electric Power Co., a big Columbus, Ohio, utility that owns coal mines and periodically does coal-by-wire transactions.[73] Such hedging would generally be used only to lay off excessive risks, because as the hedge reduces risks it also reduces the profitability of the deal that is hedged.

U.S. industry has already learned how risky complex derivatives can be. In 1993 through 1995 over a trillion dollars in derivatives value vanished from corporate balance sheets. The debacle was touched off by the sudden illiquidity of the mortgage-backed securities (MBS) market caused by an unanticipated rise in interest rates. In order to increase profits, banks invested in "tranches" of MBSs that were supposed to be less risky than corporate lending, but safer and more profitable than money market funds. How could a security that was guaranteed by a government agency, Ginnie Mae, as well as being secured by land, become valueless? Very simply, it turned out. When interest rates increased sharply, the MBS securities not only fell in value but became impossible to sell—the entire MBS market lost its liquidity. It then took banks years to gradually sell off their mortgage backed securities as interest rates fell and the market slowly recovered.

Normal corporate financial expertise apparently was not equal to the task of assessing the risk involved in mortgage-backed securities or in other kinds of complex derivatives options. Even multinational corporations like Sears Roebuck successfully sued banks for misrepresenting the risk involved in complex derivatives. Orange County, California settled a similar suit against Merrill Lynch in a case involving the bankruptcy of the county. Money market funds are supposed to invest their capital in treasury bills, notes, and bonds that have no business risk at all. But when these money market managers speculated in derivatives the funds "broke the buck" and returned less than a dollar for each dollar invested in these supposedly low risk funds.

These scenarios are relevant to the emerging markets in energy because energy markets behave much like a futures markets, and the energy futures contracts are almost always complex and risky.

An Example: Oglethorpe and LG&E

One example of a long-term energy contract that behaved much like a commodities future contract was the contract that Louisville Gas & Electric Company (LG&E) made with Oglethorpe Power to supply over half of Oglethorpe's electricity needs for fifteen years.[74] Oglethorpe entered this agreement despite the fact that they were, themselves, a large electricity generator. Oglethorpe serves 2.6 million households and supplies power to 39 electric cooperatives. Yet Oglethorpe felt that it could not match the massive markets and marketing skills of LG&E, and was willing to buy their power because it was cheaper.

Contracts like Oglethorpe's with LG&E act like puts and calls. From Oglethorpe's point of view the contract behaved like a call, because the contract for the low cost electricity to be supplied by LG&E would only become valuable if Oglethorpe's cost of electrical production went above the contract price agreed to by LG&E. This would have been the equivalent of Oglethorpe's betting that the cost of electricity would increase. LG&E was betting exactly the opposite, that the price of electricity would fall, so from LG&E's point of view, the contract behaved like a put. LG&E's profit on the contract would be the difference between what they hoped would be a higher contract price and the lower market price they expected in the future. From a production point of view, LG&E was betting that their expertise and scale of operations would let them produce electricity at a price that was less than the market—less than the price that Oglethorpe could charge for electricity.

When Oglethorpe and LG&E negotiate long-term electricity contracts like this they are speculating on the future prices of energy. The contracts are very risky. When Oglethorpe decided to let outsiders—the market—supply them with power, they entered a transaction which was much like purchasing a call on future commodity prices. Since Oglethorpe's business is producing and selling electricity rather than speculating on commodity prices, they might want to hedge the risk of their contract with LG&E if it appears later that electricity prices may fall precipitously. Oglethorpe could do this by selling a contract to supply electricity—a put. Thus, Oglethorpe's purchase contract with LG&E (a call) would be canceled out by Oglethorpe's sales contract (a put). Since both contracts depend on energy price movements, the price of the call would fall with energy prices in exactly the same amount that a put would rise if the two contracts were for equivalent amounts of electricity, so the change in put and call values would cancel out. Thus Oglethorpe would have hedged away all of the risk of electricity price movement (and lost their profits as well) as long as the two balancing contracts were in effect.

Since it will be necessary for large utilities to speculate in complex commodity options and futures, the Financial Accounting Standards Board (FASB) accounting rules for these options will become an important part of the financial skills of utility treasurers. The rules on how accounts report the value of complex derivatives appears to be changing in an important way. In July 1997, two detailed letters were sent to the FASB complaining about the Board's intention to draft a rule that would force companies to treat a change in the value of stock options as an expenses.[75] One letter was from the heads of leading financial institutions and the other letter was from William Greenspan, the chairman of the Federal Reserve Bank. The letters threatened to lobby Congress if the FASB did not back down on the proposed rule change.[76]

The Fed Approach

Greenspan called the proposed rule "a piecemeal approach" that might present misleading financial information[77] and be so burdensome that it might prevent some companies, especially smaller companies, from using options to hedge their risks.[78] Greenspan further suggested that the FASB simply expand existing types of disclosures[79] and require only larger companies to disclose the fair market value of their options.[80] Businesses feared that the new rules would convert the volatile fluctuations in options values into disturbing and misleading earnings fluctuations if these changes were reported on the income statement.

The FASB Approach

The Securities and Exchange Commission (SEC) supported the FASB rule.[81] "Financial markets demand transparency," said Michael H. Sutton, the commission's chief accountant. "Having billions of dollars of derivative transactions unaccounted for creates an intolerable risk for investors."[82] The Board's chairman, Edmund Jenkins, pointed out in his reply to Greenspan's letter, that "Your approach—when compared to ours—would reduce the information available to investors and creditors."[83]

In August 1997, the FASB announced that it planned to adopt its proposed rule on accounting for derivative securities despite the recent criticism.[84] Because the FASB intends to require publicly-held companies to adjust their earnings to reflect changes in the market value of their derivatives contracts,[85] these accounting rules will be especially important for U.S. utilities because these utilities will be accumulating increasing quantities of energy derivatives and will be required to report the changes in these derivatives' values as changes in earnings. This means that these very large utilities may soon be showing very large fluctuations in their earnings that are completely inconsistent with their traditional image as safe, sanctioned monopolies. The market's reaction to this apparent increase in corporate risk will be interesting because of the very large amounts of bond and equity offerings necessary to finance the huge investments of the capital-intensive utilities industry. If the capital costs of U.S. utilities increase, this will put a lot of financial pressure on an industry that has often had rates of return of less than 15 percent.

Optimizing Utility Investments

Many analysts thought that PacifiCorp's prize among Energy Group's diverse assets was Eastern Electricity, as opposed to their coal holdings.[86] However, Douglas Kimmelman, the head of the utility group at Goldman Sachs, said, "We clearly see this transaction as a trendsetter in spotlighting the great values that an electricity marketer can unlock from a vast base of coal supply contracts."[87] But it is now obvious that PacifiCorp's acquisition of Energy Group, PLC involved much more than the purchase of either electric plants or coal mines. The multi-faceted deal actually involved:

- Power for coal deals
- Coal-energy convergence
- Energy arbitrage
- Coal-by-wire
- Coal-tolling
- Commodity hedging
- The optimization of transportation costs

From the broadest perspective, PacifiCorp's chief executive, Frederick W. Buckman, thinks that the energy deals of the future, like PacifiCorp's acquisition of Energy Group PLC, must hope to lower prices as well as increase profits by concentrating on the optimization of global fuel procurement and management.[88]

If the deregulation of electrical utilities makes the optimization of fuel procurement management the key to competition and profitability, the companies that will do it first will be the large, specialized utilities in the top tier of the U.S. utilities industry. This is apparently already the case, because Enron's $2.1 billion merger with Portland General Corporation in July 1996,[89] and the $7.7 billion merger of Duke Power and PanEnergy the following November,[90] established the idea that the combination of gas and electrical utilities would produce a powerful entity. Enron's acquisition of Portland General caused a frenzy of bidding for natural gas companies, which increased the prices of these companies from coast to coast. CalEnergy's $1.55 billion acquisition of U.K.-based Northern Electric PLC in October 1996 pointed out that there were bargain utilities available overseas, and gave the gas-electricity merger trend an international dimension.[91]

PacifiCorp and Enron have established two major market niches that could attract other competitors. Whereas PacifiCorp is specialized in swapping coal for electricity, Enron is a specialist in swapping gas for electricity. Enron has said that it intends to stay out of PacifiCorp's coal-based niche. Despite its broad array of products, Enron's president and chief operating officer, Jeffrey K. Skilling, says PacifiCorp has staked out at least two areas that his company wants to de-emphasize—fuel procurement and ownership of utility distribution assets. "You don't want to have everybody trying to be an Enron look-alike," said Skilling.[92]

To get the cash necessary to close the deal for Energy Group, PacifiCorp will sell its big telecommunications subsidiary to Century Telephone Enterprises Inc. for $1.53 billion.[93] They might also sell their financial services business, some power plants, and issue more stock as well.[94] These deals would leave Pacifi-

Corp focused much more sharply on its core businesses in coal mines, electricity distribution, and electricity generation using hydroelectric, gas, and coal plants.[95]

NEW FORMS OF POLITICAL RISK

British Taxes on Unregulated Utilities

The previous sections show New York struggling with a state power authority that already has conflicting roles that will become even more convoluted under a deregulated power market. Although Britain's energy markets are much more deregulated than those in the United States, interesting conflicts with these free markets arose when Britain elected a Labor government in 1997. The conflicts were thrashed out over Tony Blair's imposition of a windfall profits tax on large corporations. The irony is that some of the largest corporations subject to the new windfall tax are huge utilities that have recently been acquired by U.S. utility companies who thought them to be unregulated—and thus, un-windfall taxed. Unfortunately the Labor government had to find some way to finance other reforms in their campaign platform.

The Windfall Tax on British Utilities

In May 1997, Gordon Brown, Chancellor of the Exchequer, was prepared to unveil the new Labor government's forthcoming budget, when British Telecommunications (BT) announced their pre-tax profits had increased to £3.2 billion.[96] Prime Minister Tony Blair responded by announcing his intention to levy a windfall profits tax on privatized British utilities that would raise £5 billion. Although the new revenues would be used to pay for reforms in Labor's campaign platform, BT's chairman, Sir Iain Vallance, said, "I wouldn't have voted Labor or put this government into power if BT had been mentioned in the manifesto," and added that BT said that it would contest the tax in court.[97] Had Blair specified a company or set of companies, the tax would have been deemed unlawfully discriminatory.

BT was especially disappointed, because it had helped Labor promise to connect schools and libraries to the information superhighway.[98] Now they would have to pay for doing this. The regional electrical companies planned to meet with Treasury officials, as did Southern Electric, and CalEnergy, which had recently acquired Northern Electric. CalEnergy was believed to have already made a provision of from £80 to £100 million for the tax.

It made a big difference to the utilities whether or not transportation and water utilities would be included in the tax. If they were, it would spread the levy, which, otherwise was expected to range upward from £1.3 to £1.7 billion for the RECs and from £900 to £1,350 million for the electricity generators.[99] Another point the utilities lobbied intensively was whether the tax would be levied on excess profits or would be a much higher tax based on current sales.[100]

The windfall tax proposal provided evidence that more fundamental disagreements between business and Labor might occur. Nonetheless, Brown, who ranked second in power to Blair, invited a number of businessmen to head government task forces on improving Britain's competitiveness, level of investment and long-term performance.[101] Although Brown proved himself congenial with business, Trade and Industry Secretary Margaret Beckett was expected to clamp down on City takeovers. "She is old Labor, and doesn't understand markets and competitiveness," said one businessman. "Hostile takeovers are an essential discipline on companies," said Tim Melville-Ross, director-general of the Institute of Directors. "The markets must be left to decide."[102]

Precedents for the Tax

Not everyone thought the windfall tax was a perversion of free markets. British business editorialist Simon Jenkins thought that, "from the moment that the monopoly utilities were sold without any profit drawback, a surtax was on the cards." The tax was nothing new, rather, Brown's windfall tax was "in line of descent from Nigel Lawson's negative external finance limit and Clare Spottiswoode's cash handback."[103] The Tories had taxed the sudden profits that banks made in 1991 as a result of the government's policy on interest rates, and in the late 1980s Lawson levied the "negative external finance limit" on utilities, telling them how much of their excess profits to send to the Treasury.[104]

In current terms, British gas regulator Claire Spottiswoode's May 1996 price and investment regime included a *de facto* windfall tax when she instructed British Gas to give every consumer a pound a week bonus. Stephen Littlechild, the electricity regulator, also responded in this tradition when he transformed the structure of British electricity distribution "in response to widespread public concern" at excess profits.[105]

If moral justification for the windfall tax should be necessary, Jenkins pointed out that the National Audit Office found that by 1990 £2.3 billion had been lost by undervaluations to promote the political return to privatization.[106] The subsequent bonanza of utility profits was paid for by sacking tens of thousands of workers after privatization of the utilities. "After privatization the utilities bosses pocketed obscene pay rises in defiance of public opinion." Jenkins concluded:

The Tory Government relinquished ownership of the state monopolies, but it did not relinquish control. Nicholas Ridley famously admitted that control would be easier after privatization. So labor has no need to denationalize. It can control the monopolies without ownership. A windfall tax is just one of many levers of that control.[107]

Contesting the Tax

Sir Ian Vallance of British Telecom (BT), and other privatized utilities said they would fight the windfall tax in European courts on the grounds that it was discriminatory because it was not levied equally on all utilities.[108] Under Common Market law, this would make the windfall tax just as discriminatory as preferential subsidies provided to firms in one industrial sector relative to firms in another sector which received no subsidies.[109] BT's vociferous public rejection of the tax made it virtually certain that BT would be a target of the tax, because labor would not want to back down in the face of a threat.[110] After having cuddled up to the Laborites before the election, at least five other utilities —Wessex Water, British Airports Authority, Energy Group, and Anglian Water —appeared likely to support Sir Ian's challenge.[111]

Sir Ian opened up the prospect of legal action on the day that he announced a record £3.2 billion profit for his company. The company claimed, nonetheless, that it had no excess profits and would provide no funds to cover the tax in its first-quarter accounts, because there was not enough information to establish what figure might eventually be required. The government said the tax would be used to fund a welfare-to-work program for 250,000 unemployed young people.[112]

PowerGen, one of the more recently privatized electricity generators, lobbied the government to levy the windfall tax more heavily against the utilities that were privatized and sold off in the first "tranches."[113] This would be more equitable, argued PowerGen, because the British government retained large interests in the companies sold in the later tranches, and benefited greatly from the runup in share value created by capital bidding for these last firms—which included PowerGen.[114]

PowerGen's argument would decree that the windfall tax be levied most heavily on the water companies sold in the first tranches after privatization. Over half of the returns generated by BT since privatization had already gone to taxpayers, whereas for generators, the figure was just under half this amount, with water company returns contributing only 6 percent to taxpayers. This meant that any relatively simple formula based on excess profits and not taking into account the phased sale of the privatized utilities would be inequitable.[115]

PowerGen has had a legacy of conflict with the British government since 1995, and this may partly account for their vociferous argument against the windfall profits tax. In September 1995, PowerGen, the second largest generator in the U.K. made a £2.5 billion bid for Midlands Electricity PLC which was blocked in May of 1996 because the combination was deemed anti-competitive.[116] At the same time, Ian Lang, the British Secretary of State for Trade and Industry under Prime Minister John Major, stopped National Power's £2.5 billion takeover of Southern Electric. Lang thought the mergers would have been a step toward the vertical integration of electrical generation companies in Britain, and if the mergers took place, National Power and its smaller competitor, PowerGen, would control 90 percent of the generating plants in England and Wales.[117]

Having been denied their domestic acquisitions, both National Power and PowerGen began to invest in Asia, Australia, Europe, and the United States. By 1997, National had invested £1 billion, and PowerGen £700 million abroad.[118]

Even more relevant to the windfall tax dispute was the opinion of Patricia Hodgson, Director of Policy and Planning for the British Broadcasting Corporation. Hodgson pointed out that after PowerGen had been privatized, PowerGen cut its staff in half, while privatized National Power fired two-thirds of its employees. Little of the savings from this cost cutting, she claimed, had flowed to consumers as reduced electricity rates. Instead, prices rose for the first three years after privatization.[119]

Using incredibly bad timing, five companies announced double digit dividend increases in the last week of May, 1997, only a month after the announcement of Blair's windfall profits tax.[120] Ian Byatt, the water industry regulator, attacked these increases as financially unsustainable, saying, "I'm very puzzled as to why they feel it necessary to have this huge dividend growth because it is not obvious that investors in water need to get a better return [than other Sectors]." He warned the companies that they were sending out the wrong signals. "The more you demonstrate you've done frightfully well the more you encourage people to think that a utility tax is a good idea," he said.[121]

Parliament Imposes the Tax

On July 2, 1997, the Parliament passed Blair's budget, which includes both a windfall profits tax on utilities and a big business tax cut that will give Britain's big business the lowest tax rates of any Western industrial country.[122] This signaled the continued growth of the British economy although the middle class will have to wait for tax cuts and the working class will have to wait for jobs. Brown's budget continued Tory policies by keeping the Britain's focus on private business as the key for new investments and increased profits at the expense of new welfare and social spending.[123]

The windfall profits tax fulfilled a campaign promise and its expected $7.6 billion in revenues will be used to pay for retraining workers laid off during the Conservative government's program of privatization. The budget disclosed for the first time that the windfall tax would be computed on the profits that the privatized companies made in moving from the public to the private sector, and would be paid in four installments over 1997-1998.[124]

The Labor government still appears to be pro business despite the windfall tax. The British business editoriialist Stanley Reed summed up the situation, saying that in Britain, business was starting to get the message:

The tax was dreamed up by now-Chancellor Gordon Brown's aides well before the election. Brown wanted money for his estimated $5 billion job creation program. The solution: a special tax on the water and electric companies that were privatized under the Conservatives. These companies are an easy target because the public believes the Tories sold them too cheaply and because many of their executives have done very well for themselves through buyouts and salary increases.[125]

Britain's Fear of Deregulation

The Pacificorp-Energy Group Merger

PacifiCorp's $5.8 billion in cash bid for Energy Group PLC. of Britain in June of 1997, was a merger that revealed the type of political risk a foreign utility investor might face in Britain while a new government sorts out its priorities with relation to energy policy. The merger also revealed the Oregon-based utility's desire to extend their energy trading practices into the international market. PacifiCorp's respectable, $11.30/share offer for Energy Group, was an indication of value that this coal-rich asset would have in their portfolio of energy operations.[126]

PacifiCorp's bid came late in Britain's cycle of utility deregulation and privatization. This cycle was followed by a round of foreign acquisitions of British utilities which appeared to be much better values under private management. The prices of the remaining utilities had been bid up to at least their fair market value by 1997, when Labor won their first election in years.

Despite other regulators' recommendations to quickly approve the PacifiCorp-Energy Group merger, on August 3, 1997, Margaret Beckett, the new Trade and Industry Secretary, ordered a full-scale regulatory review of PacifiCorp's planned takeover by the Monopolies and Mergers Commission. The Trade Board cited concerns about maintaining regulatory control over the merged company and other broad "public interest" questions about government supervision of the corporation rather than expressing the usual concern about specific competitive issues.[127]

Although the Trade Board's decision to investigate brought to a halt any further legal steps toward a merger, it might prove to be only a "speed bump" in PacifiCorp's bid. As evidence of this, the corporation continued with its financing plan to conclude the bid. For example, PacifiCorp proceeded with the $1.53 billion sale of its phone unit to generate cash for the deal. Beckett's decision came at just the time the new Labor government was in the process of developing its overall policy on utility mergers. Still, the disapproval of the deal meant more delays for PacifiCorp and Energy Group:[128]

- Shareholders of Energy Group will be asked again to approve the terms of the transaction.
- PacifiCorp might spend months responding to requests.
- There is the possibility of public hearings.
- The government might recommend changes in the deal.
- The government might seek to block the deal altogether.

On August 13, 1997, it became apparent that the British Trade Board's hearings on PacifiCorp's bid had cost the company at least $65 million in foreign exchange losses.[129] The company's $5.8 billion plan had offered $11.30 per share for Energy Group, so PacifiCorp had purchased options giving it the right to buy about £1.35 billion at an effective exchange rate of $1.63. In an attempt to cut its hedging costs, the options later were converted to forward purchase commitments, which required PacifiCorp to make the future currency transactions. The Trade Board's decision to technically terminate PacifiCorp's bid pending the Monopolies Commission's investigation made that purchase inadvisable, although the bid might be renewed later, after the investigation. Unfortunately, the exchange rate fell to $1.58 for each pound sterling, causing PacifiCorp an unhedged loss of $56 million.[130]

NOTES

1. Carlos Tejada, "NGC to Acquire Destec for $127 Billion; Natural-Gas Concern Aims to Stake Out Position in Electicity Market, " *Wall Street Journal*, February 19, 1997, pp. A1, A2.

2. Agis Salpukas, "$1.9 Billion Hostile Bid for Utility," *New York Times*, July 16, 1997, pp. D1, D18.

3. *Ibid.*

4. Salpukas, "$1.9 Billion Hostile Bid," pp. D1, D18.

5. *Ibid.*

6. James P. Miller and Steven Lipin, "CalEnergy Launches Another Hostile Bid," *Wall Street Journal*, July 16, 1997, pp. A1, A3, A4.

7. Salpukas, "$1.9 Billion Hostile Bid for Utility," pp. D1, D18.

8. Richard Perez-Pena, "Rate Cut Questions," *New York Times*, March 16, 1996, p. 42; Jeff Bailey, "Niagara Mohawk Plan is a Small Step Toward Easing Utilities' Power Woes," *Wall Street Journal*, March 12, 1997, pp. A1, A4; Agis Salpukas, "Utility Seeks to End Costly Pacts with Power Suppliers," *New York Times*, March 11, 1997, pp. D1, B8; and Agis Salpukas, "Niagara Deal with Independents Could Reduce Price of Electricity," *New York Times*, July 11, 1997, pp. D1, B5.

9. Salpukas, "$1.9 Billion Hostile Bid for Utility," pp. D1, D18.

10. *Ibid.*

11. Miller and Lipin, "CalEnergy Launches Another," pp. A3, A4..

12. *Ibid.*

13. Salpukas, "$1 9 Billion Hostile Bid," pp. D1, D18.

14. Miller and Lipin, "CalEnergy Launches Another," pp. A3, A4..

15. Salpukas, "$1.9 Billion Hostile Bid," pp. D1, D18..

16. Miller and Lipin, "CalEnergy Launches Another," pp. A3, A4.

17. *Ibid.*

18. *Ibid.*

19. James Miller, "CalEnergy Ends $1.92 Billion Bid," *Wall Street Journal*, August 18, 1997, pp. A1, A4.

20. *Ibid.*

21. Salpukas, "$1.9 Billion Hostile Bid," pp. D1, D18.

22. *Ibid.*

23. Dana Canedy, "Suitor Drops $1.9 Billion Bid for Utility," *New York Times*, August 16, 1997, pp. 35-36

24. Miller and Lipin, "CalEnergy Launches Another," pp. A3, A4.

25. Canedy, "Suitor Drops $1.9 Billion," pp. 35-36.

26. Miller and Lipin, "CalEnergy Launches Another," pp. A3, A4.

27. Canedy, "Suitor Drops $1.9 Billion," pp. 35-36; and Miller, "CalEnergy Ends $1.92 Billion Bid," pp. A1, A4.

28. Miller and Lipin, "CalEnergy Launches Another," pp. A3, A4.

29. Canedy, "Suitor Drops $1.9 Billion," pp. 35-36.

30. Miller, "CalEnergy Ends $1.92 Billion Bid," pp. A1, A4.

31. Canedy, "Suitor Drops $1.9 Billion," pp. 35-36.

32. *Ibid.*; and Miller, "CalEnergy Ends $1.92 Billion Bid," pp. A1, A4.

33. *Ibid*, and Canedy, "Suitor Drops $1.9 Billion," pp. 35-36.

34. The material for this section comes from an article by Agis Salpukas, "When Electricity Goes Private, Deregulation May Change New York Power Authority," *New York Times*, July 11, 1997, pp. D1, D2, D3.

35. Salpukas, "$1.9 Billion Hostile Bid," pp. D1, D18; and Miller and Lipin, "Cal-Energy Launches Another," pp. A3, A4.

36. Agis Salpukas, "When Electricity Goes Private, Deregulation May Change New York Power Authority," *New York Times*, July 11, 1997, pp. D1, D2, D3.

37. John Doukas, "The Effect of Corporate Multinationalism on Shareholders' Wealth: Evidence from International Acquisitions," *Journal of Finance*, December 1988, 43(5), pp. 1181-1175.

38. "Southern Co.: Plan to Purchase Utility Dropped after UK Action," *Wall Street Journal*, May 9, 1996, p. B4.

39. "Britain's Electricity Shocker," *Economist*, April 13, 1996, 339(7961), pp 14.

40. "Entergy Will Pay $2.1 Billion for a Utility, London Electricity," *New York Times*, December 19, 1996, pp. D1, D5; and "Dominion Agrees to Buy Power Utility," *New York Times*, November 14, 1996, pp. D1, D4.

41. *Ibid.*

42. Agis Salpukas, "CalEnergy Offers to Buy British Utility," *New York Times*, October, 19, 1996, D1, D7; and "British Utility Lifts Payout in Face of Bid," *New York Times*, December 11, 1996, p. D6.

43. "British Utility Lifts Payout," p. D6.

44. Salpukas, "CalEnergy Offers to Buy," D1, D7; and "British Utility Lifts Payout," p. D6.

45. "CalEnergy Bid for Utility is Victorious; British Distributor is Latest Acquisition," *New York Times*, December 25, 1996, pp. D1, D3.

46. Rudi Dornbusch, "Economics Viewpoint: The Asian Juggernaut isn't Really Slowing Down Dynamism," *Business Week*, July 14, 1997, p. 16.

47. Agis Salpukas, "Big U.S. Utility Spreads Its Reach to Berlin," *New York Times*, May 24, 1997, p. D2.

48. Simon Holberton, "Generators Plug in Abroad," *Financial Times*, May 26, 1997, p. 17.

49. "Southern Co. Unit Getting Role in Asia," *New York Times*, October 10, 1996, p. D4.

50. Dornbusch, "The Asian Juggernaut," p. 16.

51. Brookings Instititution papers by Barry Bosworth and Susan Collins, cited in Rudiger Dornbusch, "Economics Viewpoint: The Asian Juggernaut isn't Really Slowing Down Dynamism," *Business Week*, July 14, 1997, p. 16.

52. Dornbusch, "The Asian Juggernaut," p. 16.

53. David L. Kaserman, "The Measurement of Vertical Economics and the Efficient Structure of the Electric Utility Industry," *Journal of Industrial Economics*, September 1991, 39(5), pp. 483-502.

54. Salpukas, "Big U.S. Utility Spreads," p. D2.

55. *Ibid.*

56. *Ibid.*, and Peter Fritsch and Maureen Kline (Milan), "Enron, Italian Utility ENEL Expected to Announce Power-Marketing Venture," *Wall Street Journal*, June 3, 1997, pp. A1, A3.

57. Salpukas, "Big U.S. Utility Spreads," p. D2.

58. *Ibid.*

59. Michael Sayers, "The Death of Protected Profits," Morgan Stanley, International Investment Research, September 26, 1996, pp. ___, cited in Judith B. Sack and Robert L. Chewning, "Global Electricity Strategy: My Two Cents Worth (Or the Sustainable Price of Power)," Morgan Stanley, International Investment Research, February 6, 1997, p. 1.

60. Agis Salpukas, "Big U.S. Utility Spreads," p. D2.

61. Benjamin A. Holden, "UtiliCorp and Peco, Aided by AT&T, To Launch One-Stop Utility Service," *Wall Street Journal*, July 24, 1997, pp. A1, A3.

62. Agis Salpukas, "Pacificorp is Said to Reach Deal to Buy British Utility," *New York Times*, June 12, 1997, pp. D1, D7.

63. *Ibid.*

64. *Ibid.*

65. Benjamin A. Holden, "PacifiCorp Pursues Energy-Swap Plans," *Wall Street Journal*, June 16, 1997, pp. A1, B4.

66. Gary McWilliams, "Enron's Pipeline into the Future," *Business Week*, December 2, 1996, p. 82.

67. Holden, "PacifiCorp Pursues Energy-Swap Plans," pp. A1, A3.

68. *Ibid.*

69. *Ibid.*

70. *Ibid.*

71. Holden, "PacifiCorp Pursues Energy-Swap Plans," pp. A1, A3.

72. *Ibid.*

73. *Ibid.*

74. *Ibid.*

75. Agis Salpukas, "Utility Seeks Partner to Aid Power Needs," *New York Times*, November 20, 1996, pp. D1, D2.

76. Floyd Norris, "Greenspan Opposes Accounting Change on Derivatives," *New York Times*, August 7, 1997, pp. D1, D8.

77. Floyd Norris, "Market Place: Some Big Financial Guns Attack a Proposal on Derivatives by the Accounting Standards Board," *New York Times*, August 1, 1997, pp. D1, D6.

78. Norris, "Greenspan Opposes Accounting," pp. D1, D8.

79. "Accounting Board to Adopt Derivative Rules," *New York Times*, August 12, 1997, pp. D1, D2; and Elizabeth MacDonald and Stephen E. Frank, "FASB Rejects Fed Chairman's Request To Soften Proposed Rule on Derivatives," *Wall Street Journal*, August 12, 1997, pp. A1, A2, A9.

80. "Accounting Board to Adopt," pp. D1, D2.

81. MacDonald and Frank, "FASB Rejects Fed Chairman," pp. A1, A2, A9.

82. "Accounting Board to Adopt," pp. D1, D2.

83. Norris, "Greenspan Opposes Accounting," pp. D1, D8.

84. "Accounting Board to Adopt," pp. D1, D2.

85. *Ibid.*

86. MacDonald and Frank, "FASB Rejects Fed Chairman," pp. A1, A2, A9.

87. Agis Salpukas, "Says Pacificorp Considers a Bid," *New York Times*, June 11, 1997, pp. D1, D2; Salpukas, "Pacificorp is Said to Reach Deal," D1, D2; and Holden, "PacifiCorp Pursues Energy-Swap," pp. A1, B4.

88. Holden, "PacifiCorp Pursues Energy-Swap Plans," pp. A1, B4.

89. *Ibid.*

90. Benjamin A. Holden, "Enron Corp. Has Accord to Buy Portland General," *Wall Street Journal*, July 22, 1996, p. A3; Sullivan Allanna, "Enron Deal Signals Trend in Utilities," *Wall Street Journal*, July 23, 1996, p. A3; and Allen R. Myerson, "Enron Will Buy Oregon Utility in Deal Valued at $2.1 Billion," *New York Times*, July 23, 1996, p. D1.

91. Steven Lipin and Peter Fritsch, "Duke Power Plans to Acquire PanEnergy in Stock Transaction of About $7.7 Billion," *Wall Street Journal*, November 25, 1996, pp. A1, A3.

92. Salpukas, "CalEnergy Offers to Buy," pp. D1, D7; "British Utility Lifts Payout," p. D6; and "CalEnergy Bid for Utility is Victorious.," pp. D1, D3.

93. *Ibid.*

94. *Ibid*, and Benjamin A. Holden, "PacifiCorp is Likely to Finance Purchase in U.K. with Sale of Telecom Business," *Wall Street Journal*, June 17, 1997, p. B4.

95. Salpukas, "Pacificorp is Said to Reach Deal,"pp. D1, D2,

96. Holden, "PacifiCorp Pursues Energy-Swap," pp. A1, B4.

97. "Tax Row Mars Labor's Honeymoon," *Times*, May 18, 1997, p. 5.

98. *Ibid.*

99. *Ibid.*

100. *Ibid.*

101. *Ibid.*

102. *Ibid.*

103. *Ibid.*

104. Simon Jenkins, "Monopoly Game Over," *Times*, May 21, 1997, p. 24.

105. *Ibid.*

106. *Ibid.*

107. *Ibid.*

108. *Ibid.*

109. Philip Webster and Christine Buckley, "Windfall Tax Faces Legal Challenge," *Times*, May 16, 1997, p. 1.

110. *Ibid.*

111. *Ibid.*

112. *Ibid.*

113. *Ibid.*

114. David Wighton, "Utilities Call for Equity in Windfall Tax," *Times*, May 20, 1997, p. 12.

115. *Ibid.*

116. *Ibid.*

117. "Britain to Review Utility Acquisitions," *New York Times*, November 24, 1996, p. D12; "Powergen of Britain May Bid for Midlands," *New York Times*, September 16, 1995, pp. 18, 32; and Tara Parker-Pope, "UK Utilities Generate Takeover Frenzy; Latest

Bid, Powergen's for Midlands, Faces Hurdles," *Wall Street Journal*, September 18, 1995, pp. A16, A14.

118. Stephanie Strom, "International Business: British Reject 2 Power-Industry Takeovers; Utility Stocks Up on Merger Speculation, Fall Sharply in London," *New York Times*, April 25, 1996, p. D7.

119. Simon Holberton, "Generators Plug in Abroad," *Financial Times*, May 26, 1997, p. 17.

120. Strom, "British Reject 2 Takeovers," p. D7.

121. Leyla Boulton, "Regulator Slams 'Huge' Water Dividend Rises," *Financial Times*, June 4, 1997, p. 8.

122. *Ibid.*

123. Youssef M. Ibrahim, "Blair Gains Tax Cut for Business, But the Rest of Britain Must Wait," *New York Times*, July 3, 1997, pp. A1, A5.

124. *Ibid.*

125. *Ibid.*

126. Stanley Reed, "Britain: Business Starts to Get the Message," *Business Week*, June 2, 1997, p. 58.

127. Salpukas, "Pacificorp is Said to Reach Deal," pp. D1, D2.

128. Benjamin A. Holden and Helene Cooper, "Britain to Conduct Antitrust Review of PacifiCorp Deal," *Wall Street Journal*, August 4, 1997, pp. A1, B5.

129. *Ibid.*

130. Benjamin A. Holden, "PacifiCorp to take $65 Million Charge on Currency Deals," *Wall Street Journal*, August 13, 1997, pp. A1, A6.

Chapter 4

The U.S. Antitrust Environment for Utility Mergers

THE CHANGING ANTITRUST ENVIRONMENT

Reaganomics; the Republican majority in Congress; and a new, middle-of-the-road Clinton administration have provided the political background for a U.S. antitrust policy environment that is more permissive to mergers than it formerly was. The Federal Trade Commission (FTC) is processing more antitrust cases than the antitrust division of the U.S. Justice Department and, for the first time, administrative law is perhaps becoming more important in setting antitrust policy than case law. FERC has followed the lead of the Justice Department and the FTC by adopting new, streamlined rules designed to cut down on time-consuming, expensive courtroom battles and extensive, technical factual proofs of market power and competitiveness. These antitrust traditions have yielded in favor of administrative checklists designed to provide a quick decision for or against utilities mergers.

The Current Evolution of Antitrust Regulation

In 1996, Joel Klein, a Harvard-educated long-time Washingtonian, became head of the Justice Department's antitrust division, taking over the reins of U.S. antitrust policy from Anne K. Bingaman.[1] Klein announced a simple agenda: to reduce the government's regulation of business while working for increased international cooperation to stamp out global cartels.[2] Klein has been attempting to reduce regulation in order to increase market competition. He has also tried to clarify the legal boundaries in antitrust cases involving intellectual property, health care, and high tech industry.[3] This minimalist policy made Klein look less like a crusader than Bingaman.

Klein's policies contrast with those of William Baxter, the Reagan administration's free-market-minded antitrust chief. Baxter codified the idea that merger policy was a numbers game framed by statistics defining market share. He wrote this philosophy into guidelines permitting all mergers that did not concentrate the market beyond a predetermined limit. The result was a virtual free-for-all.[4] In one sense this free-for-all has never ceased, because although mergers still occur in waves, they have occurred in such great numbers over the last fifteen years that the merger phenomenon is now considered more a method of capital investment, than an extraordinary occurrence. Mergers annually account for approximately 20 percent of all capital investment.

The Federal Trade Commission

Besides the antitrust division of the Justice Department, the other federal agency most influential in antitrust disputes is the FTC. The FTC investigates situations involving antitrust violations and practices in restraint of trade and adjudicates these situations administratively. These proceedings make the FTC an important mediator in many antitrust decisions. Many of the FTC's antitrust investigations arise as a result of the complaints of competitors of unfair trade practices.

Robert Pitofsky, the head of the FTC in 1998, was a one-time believer in the big-is-bad doctrine, but his work at the Trade Commission has largely eliminated the discontinuity between traditional antitrust policy and modern economics.[5] The key question for Pitofsky's FTC staff is how will a merger affect costs? And how will those costs affect prices?[6] These are questions that previously concerned economists more than antitrust officials.

Now that the FTC's antitrust policy has been refined by questions of economic efficiency, questions of costs and prices must be answered separately for each product and each market. The goal of staff research now is to estimate whether the increased concentration caused by a merger will prevent the combined companies from lowering prices or give them so much market power that this is unlikely.[7] If the result of the merger causes greater efficiency and lower prices, the FTC is likely to approve mergers of even very large corporations. Such was the case of the mergers of, first, Manufacturer's Hanover Bank with Chemical Bank, and then the merger of Chemical Bank with the Chase Manhattan Bank—all mergers of the world's largest banks within a period of five years.[8]

Mergers of Different Segments of the Same Market

The new antitrust rules also allow the merger of large companies specializing in different segments of the same market. For example, in the financial services industry, Dean Witter, a retail broker, merged with Morgan Stanley, an investment banker. This same trend can be seen in the utilities industry, epitomized by the merger of Enron with Portland General Corporation. The combination of Enron, a Texas-based natural gas marketing giant, with Portland General, a large

Oregon-based electrical utility, set the stage for Enron's national energy marketing effort, which incorporated the idea of gas-electricity convergence.

There appears to be no obvious antitrust problem with this merger because Enron and Portland General are from different segments of the energy market, one from gas, the other from electricity. The utilities are also from different parts of the country, so there are no obvious regional monopolies involved. Although this partial analysis fits the general trend of antitrust regulations favorably, the merger was consummated specifically to build a giant firm so large that it would be able to market energy nationwide.

The merger was based on the new concept of energy convergence, which postulates that one company can sell energy equivalents of the energy source they produce the cheapest wherever they can get the best price. In the case of Enron and Portland, the convergence was between electricity and gas. The new, low-cost technologies of gas-fired generation and the market technology of selling interchangeable energy equivalents of gas and electricity could be seen welding the gas and electrical utility industries into one.

The technologies of energy convergence at work in the Enron-Portland merger were viewed by firms in the utilities industry as having created a small merger in one industry rather than a large one in two industries, and the regulators also saw it this way. There were at least three reasons for this regulatory acquiescence. First, Enron and Portland General introduced the concept of electric companies competing in the gas industry and vice versa, which would produce more competition and please FERC. Until the Enron-Portland General merger, electric companies had thought themselves immune from takeover by outside suitors for over 60 years.[9]

The second and best reason for FERC's acquiescence, however, was that the Enron-Portland General combination was designed to sell whichever source of power was the cheapest, gas or electricity, and this created an efficient, low-cost producer. Economists among the new antitrust regulators believe that this will produce lower prices if it is done under competitive conditions. The third reason for FTC approval was that even though the merger was large, Enron-Portland were expected to compete nationwide with local utility monopolists and thereby increase competition in numerous places where competition previously did not exist. Even if Enron-Portland was a large merger in a new industry, that industry was huge because technology had just welded two other giant industries—gas and electricity—together.

From a technical and practical point of view, the determining factor in the Enron-Portland General merger was that the resulting merger provided a more efficient energy producer that could sell power at lower prices. The technical issue of whether the two large companies were in the same industry and product was less important. The question of whether mergers such as the Enron-Portland General merger implied more federal micromanagement of mergers in the new industries created by technology and marketing was considered academic compared to the practical results of creating a more efficient, competitive firm selling at lower prices.[10]

Mergers to Dominate a Small Segment of a Large Market

The Staples-Office Depot deal was not as lucky as the Enron-Portland General merger. Together, the Staples and Office Depot chains sold only 4 percent of the $170 billion worth of office supplies in America—a smaller market share than Wal-Mart. By offering service and pricing previously unavailable to small-business purchasers, these two office superstores had created a new market. As a consequence, the FTC decided that most of their customers would have nowhere better to turn if the merged office-supply chains took advantage of their new market position and raised prices by as much as 15 percent.[11] Adding to the difficulty of the companies was the discovery of an internal memo by the FTC that indicated that Staples' $4 billion offer for Office Depot was intended to stop a decline in their profit margins by removing their most aggressive competitor by merging with it.[12]

The FTC also blocked the merger of Rite-Aid and Revco for similar reasons. Rite-Aid and Revco would not have controlled a very large proportion of the total market for prescription drugs, but the commission was concerned that the merger would have enough market power to mark up the prices of pharmaceuticals millions of customers claiming under medical health insurance plans.[13]

Contiguous Utilities

The new antitrust rules that prevent mergers in narrow segments of large markets tend to work against mergers of large, contiguous utilities. Regulatory authorities are especially strict when the contiguous utilities are electrical utilities which would end up with a monopoly on the power transmission lines in the combined territories. For this reason Wisconsin Energy and Northern States Power were forced to call off their merger.[14] Wisconsin and Northern States had even planned to appoint an independent administrator for their power lines, but this was not sufficient to reassure the FERC that the combination would not be left with excessive market power.

The monopolies created by the power lines of merging contiguous utilities has caused enough concern that FERC has pioneered new ways to combat the problem presented by these monopolies. One approach allows local municipalities to condemn the power lines connected to their city's electricity customers. This approach allows towns and cities to set up their own low-cost municipal utilities to compete with larger, high-cost utilities which might have a monopoly on electricity lines. The municipalities cannot simply take the larger company's power lines, because the larger utility owns them. The lines must be condemned in a legal proceeding in which a jury may ultimately have to determine a fair market value for the lines. The municipal utility would then be able to purchase the lines and, if it were economically feasible to do so, proceed to sell lower cost municipal power over those lines to city customers. Without such a procedure there would be no way for a municipal utility to compete with a larger utility that owned the transmission lines to their city. Because the fair market value of the transmission lines is uncertain and the management of the fledgling municipal utility untested, there is a good deal of risk in assuming that the smaller utility will be able to supply cheaper power as dependably as the larger utility could.

Nonetheless, FERC's deregulatory policies encourage this kind of competition hoping that the competition will result in lower power rates. It is difficult to see how FERC can be wrong in opening up new power alternatives like this.

A second approach is the appointment of an independent administrator for the transmission lines of the merging utilities. The independent administrator would not only make certain that competitive utilities had appropriate access to the transmission lines, but would also assure that peak loads were handled effectively and that the transmission system was safe and dependable. But as the Wisconsin-Northern States merger illustrates, this alternative has not always satisfied regulators that a transmission monopoly with sufficient power to squelch competition might not exist.

The history of utilities technology has shown a number of shifts in the trade-offs between administrative and technical efficiency leading us to surmise that in the future it might be possible that technical innovations may provide new ways for utilities to share the same transmission lines without the administrative difficulties currently posed by condemnation hearings and the appointment of independent administrators.

The Transformation of the Regulators

The shift from big-is-bad antitrust rules to the efficiency arguments of economists has lead to messy disputes over the details in mergers where dueling economists fight for the technical high ground.[15] Some observers of these duels have concluded that the antitrust authorities have been transforming themselves from trust-busting policemen into regulators. This would be a perverse consequence because the new economic policies of antitrust that have pushed the marketization of deregulated utilities are driven by the idea that markets can make decisions more efficiently than people.[16]

It is difficult for the Justice Department and the FTC to avoid making these complicated factual determinations if they must estimate future prices. Under the new approach, the agencies must identify the anticompetitive aspects of the merger, then negotiate spin-offs, set pricing rules, or make reporting requirements that offset their concern for the public welfare.[17] The bargaining power of corporations in these proceedings is very limited, so the process has resulted in lots of negotiated settlements and few court cases and also in a perceived need to streamline antitrust review procedures even more.[18] One suggestion is that to simplify merger adjudications, the antitrust regulators could set formal thresholds for government intervention: if a merger would raise prices by less, say, than 5 percent, regulators would subscribe to the presumption that the social virtue of efficiency gains of the merger outweighed its vices.[19]

In early 1997, the government revised the rules that composed a crucial part of the 1992 federal merger guidelines.[20] The new rules were released jointly by the FTC and the Justice Department. These rules made allowances for weighing cost savings and other efficiencies of combinations in evaluating mergers, and made approval more likely for some corporate combinations that might otherwise have been rejected on antitrust grounds. The rules allowed companies to argue that a merger that would result in lower prices, or improved products or services, could offset other anticompetitive concerns.[21] FTC Chairman Robert Pitofsky said, "It won't change the result in a large number of cases, rather it will have the greatest impact in a transaction where the potential anticompetitive problem is modest and efficiencies that would be created are great."[22] The revised FTC and Justice Department Merger guidelines would:

- Let companies argue to the government that cost savings and related benefits might offset potential anticompetitive effects of a merger. For example, would a merger that raises anticompetitive concerns nevertheless lead to lower prices, better products or better service?
- Define which cost savings and other benefits are directly attributable to the merger itself. Would they still get these efficiencies without merging?
- Clarify what companies must do to show that cost savings will result—a road map for making their case before the government.
- Bring U.S.. antitrust guidelines more into line with those of other countries.[23]

The approach outlined in the guidelines has been reflected in FERC's approval of mergers of very large, multinational U.S. utilities, such as Pacific Gas and Electric, and Enron and the Southern Company. The reason that these large companies have been allowed to merge to their present sizes is that they have begun to market their gas and electricity nationally, and in some cases, globally. To FERC this means that as states decontrol their local gas and electricity monopolies, Enron, PG&E, and Southern are expected to enter those markets to compete with *local* companies by bidding down power rates state by state.

Combining Operations to Reduce Costs

Under the new rules, slashing overhead costs is not as significant as being able to prove that you can cut the per-unit cost of production, so the industries most likely to benefit are those with high fixed costs. For example, merging companies could argue that by combining two factories—each operating at less than full capacity—they would realize efficiencies that could lead to lower prices. Service industries, with little capital investment and high labor-usage rates, will find it difficult to combine their way to lower unit costs.[24] This means that the new rules will have their greatest effect in the manufacturing, heath care and defense industries, where there is overcapacity and high fixed costs.[25]

These rules might also favor utilities if it were possible to combine generating plants in the same way that manufacturing operations can be integrated. But this is not so. Most power plants are stand-alone operations which have been built into a dam, or built around some other source of low-cost fuel to supply customers in a surrounding territory. Although electrical utilities now sell electricity nationally off the power grid, this only solves one problem of locality: it tends to snip the connection between the power plant and its local customer area as areas of the grid with high electrical potential offer energy in oversupply to areas of lower potential at a discount. But the equipment of the power plants cannot be combined effectively with that of other plants to reduce costs in the same way that some manufacturing operations can. There is simply no way to combine two hydroelectric turbines into one larger, more efficient turbine without rebuilding both turbines.

Market Power Is Still Important

In following the lead of the Justice Department and the FTC, FERC has crafted rules that set the stage for allowing more utilities mergers by allowing the mergers of huge, non-contiguous national power marketers. These larger national marketers have been allowed to enter many states and compete with local utilities that used to operate as state-sanctioned monopolies. By setting up all of this national-local competition, FERC has dramatically reduced the potential for anti-competitive effects. This clears the way for more merger approvals that are expected to lead to larger utilities with more efficient power rates. If the combination results in a monopoly on transmission lines, however, all bets are off.

Having said this, lower prices are not everything. Classical predatory pricing reduces prices to drive out competition. Once the competition has been removed, the remaining firms are able to raise prices to monopolistic levels because without the competition, they have the power to do so. Turning this argument ninety degrees, regulators argue that utilities should not be allowed to gain excessive market power through mergers merely because they argue that the merger will produce lower prices. Without regulation, those low merger prices might well be temporary if the unit producing them gains the market power to increase the prices later. According to FTC Chairman Robert Pitofsky, "Efficiencies are most likely to make a difference in merger analysis when the adverse competitive effects, absent the efficiencies, are not great. Efficiencies almost never justify a merger that leads to a monopoly or near monopoly."[26]

Consent Decrees

One feature of the new antitrust approach to mergers has been a dramatic increase in consent decrees. These decrees moved much of the antitrust negotiations between merging companies and the regulators out of court and into administrative hearings. The decrees fit in with the administrative approach of the FTC, while they tread on the legal traditions of the Justice Department. The consent decrees appear to save companies time and legal fees, but with them come a new set of administrative headaches.

Consent decrees became necessary when the government increased the reporting requirements of companies involved in mergers. One law partially responsible for the increase was the Hart-Scott-Rodino Act of 1976. This act required companies planning mergers to report these plans to alert the Justice Department and the FTC of their intentions. It also required the companies to give the government any information that they requested, and created enormous reporting requirements which often took months to satisfy.[27] Although the requirements of the act were onerous, one result of its provisions was to give antitrust authorities greater leverage. This leverage is often used to force the parties to antitrust disputes to clarify their issues before lawsuits are filed to block the mergers.

The Hart-Scott-Rodino Act probably reduced litigation and led to ground rules that facilitated negotiations between the government and the companies. These negotiations led to consent decrees rather than law suits, and the antitrust agencies turned themselves into the regulatory monitors of these decrees. FTC commissioner Roscoe Starek III complains that the FTC is ill-suited to a role that seems to him to be one of *de facto* price regulator.[28]

The antitrust rules that turned authorities into regulators rather than policemen did not affect utilities very much, because, historically, U.S. utilities were already monitored by their local utility commissions. Since it is easy to monitor electricity rates, complicated consent decrees were not necessary.

Now that state power commissions are being abolished as fast as FERC can encourage deregulation, one possible outcome may be that consumer groups might substitute antitrust monitoring for that of the previous state power commission. This would allow the courts to monitor utilities much like other merging industries—under the provisions of consent decrees. This probably would not occur frequently, nor would it often be burdensome, because the issues in utility mergers are relatively simple—Do the utility rates go down?

In industries like pharmaceuticals, however, antitrust monitoring has been bothersome. Eli Lilly & Co. acquired PCS Health Systems hoping to improve their marketing to health maintenance organizations (HMOs). The government approved the $4 billion acquisition but imposed complex rules that set up a "fire wall" between Lilly and PCS and established complicated reporting requirements.[29] Lilly's management was not allowed to speak to managers at PCS about either pricing or contracts with suppliers, because officials did not want Lilly to benefit unfairly from knowing the terms and prices that other drug makers had offered through PCS.[30] The arrangement was burdensome because the government felt the need to assure that the companies lived up to the promises made in their consent decree to protect the public from unfair price competition by Lilly.[31] Without these assurances, Lilly would not have been allowed to merge with PCS.

Although the number of FTC consent decrees has doubled over the past five years, Pitofsky says that the FTC is mischaracterized as "Big Brother."[32] Neither the Justice Department nor the FTC can afford constant monitoring, so the agencies have to rely heavily on the complaints of competitors to verify that corporations abide by the provisions of the decrees.[33]

Unlikely Mergers

The use of consent decrees sometimes results in settlements that let mergers proceed that in the past might have been unlikely to win approval. If all of the government's objections to a merger are countered in a consent decree, antitrust officials often will let the merger proceed. One such case was the $3.4 billion buyout of West Publishing Co. by Thomson Corporation. This combination resulted in the merger of the two largest publishers of state and federal statutes and lawsuits and seemed to violate antitrust law. But the Justice Department negotiated a consent decree providing for competition. According to the agreement, the two companies were required to divest themselves of units providing overlapping services, to make some other minor changes, and when they did these things the government then said "yes."[34]

Unique Solutions

It is often surprising what the government will require in order to preserve competition. It is also surprising what a company will agree to in return for FTC approval. In 1995, in order to get FTC approval to buy German manufacturer Leybold AG, one of its competitors, Swiss conglomerate Oerlikon-Buhrle Holding AG entered a consent decree promising to sell two businesses, including their Pfeiffer subsidiary, which manufactured high-tech equipment for compact disk and semiconductor makers.[35]

Oerlikon became stymied by FTC requirements and filing deadlines and could not find a buyer for their Pfeiffer unit, so they spun it off as Pfeiffer Vacuum Technologies in the first initial public offering (IPO) ever approved by the FTC.[36] The consent decree for the spinoff was bizarre. After having second-guessed Oerlikon in every step of the IPO, the FTC required Oerlikon to pay its new rival's legal fees, sell all of its stock in Pfeiffer, and let the spun-off corporation hire away 250 of their own sales people. "We were able to cherry-pick from among their best employees, and they weren't allowed to stop us," says Wolfgang Dondorf, chief executive of Pfeiffer.[37]

FERC'S NEW RULES FOR UTILITY MERGERS

In its capacity as monitor of utilities mergers, FERC has crafted its own rules for utilities mergers that reflect the general trends in the less specialized mergers handled by the Justice Department's Antitrust Division and the FTC. The result of FERC's continuing deliberations on utilities mergers and acquisitions is a body of decisions interpreting their streamlined rules designed especially for mergers and acquisitions in the utilities arena. FERC is willing to approve a merger if it fosters competition, encourages efficiency, and does not create undue market power. Some clues to this policy gleaned from recent FERC decisions are:

- Mergers between very large utilities that are not contiguous are generally approved because they provide efficient companies that can compete in multiple regions without creating new monopolies.
- Mergers between small contiguous utilities are generally approved if they create more efficient companies.
- Mergers between very large contiguous utilities may not be approved because of the creation of excessive market power.
- If the merger of contiguous utilities involves transmission lines, FERC may require that the merger plan provide for an independent manager of the transmission lines to prevent a monopoly on power delivery.

- Sometimes even an independent operator for the transmission lines of merging contiguous utilities will not save the merger from a charge of excessive market control.
- FERC has no objections to mergers uniting electricity generators with natural gas sellers.
- If a merger might create a limited monopoly this can sometimes be solved if one or both of the merger partners divests themselves of some of the units creating the monopoly or by allowing national power marketers to enter state markets.

ACCOUNTING CHANGES

FASB Considers an End to Poolings of Interest

Because of the billions of dollars of mergers among U.S. utilities, almost any change the FASB makes is relevant to utility share prices and to their mergers. Recently, however, FASB attempted to do away with the most popular accounting method used in mergers: the pooling of interests. This move has great implications for the utilities industry simply because of the volume of its merger activity.

In 1997, the FASB considered restricting or abolishing the pooling of interests method of merger accounting in favor of the purchase method of accounting. Unfortunately, the pooling of interests method was by far the most popular accounting method in use for big stock mergers of all kinds. This method was preferred because it did not depress the future cash flows of merged corporations. The proof of the method's popularity is in the statistics: since 1992 there had been 357 poolings versus only 36 purchase acquisitions in acquisitions worth more than $100 million.[38] After the FASB's intentions were widely criticized, industry felt relieved when the FASB took a few steps back from their strong initial position favoring the purchase method of accounting.

The FASB does not like the pooling of interest method because it does not explicitly state the amount paid for the acquired company.[39] Under a pooling merger you must hunt for this information in the statement footnotes.[40] The pooling method also allows different treatments for identical mergers and acquisitions made within the United States, and the FASB has a history of pushing for uniform accounting practices. The FASB also dislikes the pooling method because it is not frequently used in international mergers and acquisitions accounting.[41]

In the short run, a shift to the purchase method of accounting was so unpopular that it was thought that it might perversely stimulate mergers. Joel Cohen, co-head of Donaldson, Lufkin Jenrette's New York mergers and acquisitions division thought that, "The biggest risk is the uncertainty. If executives aren't sure if [the rule is] going to be adopted or not, they're going to rush to get things done before any possible effective date so that you may see an acceleration in the short-term of the merger boom."[42] If Donaldson was right, this merger boom might have lasted until 1999, the proposed effective date of the new accounting rule.

Many experts said, however, that the FASB's shift from the pooling of interest to the purchase accounting method would have a long-term chilling effect on future combinations. "If the limits pass, it would not only slow down record M&A activity, but [in the long run] a third of all mergers may get torpedoed," predicted Robert Willens, a managing director at Lehman Brothers in New York.[43]

These arguments were influential in convincing the FASB not to shift to the purchase method. The fact that they yielded to the objections from the market for corporate control obliquely showed how important mergers had become as a basic method of capital budgeting, and, from a regulatory point of view, how important mergers were thought to be in promoting industrial efficiency and lower prices.

The problem with the purchase method of accounting is that it tends to reduce the reported income of the corporation after the merger and this depresses the share value of the acquiring company. This happens because under the purchase method, the acquisition's assets are recorded at their actual purchase value. This purchase value generally represents a winning bid for the target company made above the current market price of the target. The bid is generally made above the target's market price either to win a bidding contest for the target, or to freeze out other bids. So the difference between the market value of the acquisition (lower) and the purchase price (higher) is recorded as an intangible asset called goodwill. This goodwill represents the premium that was paid over market value to acquire the target company, and must be amortized over the life of the acquisition by charging off periodical amounts against corporate income. Deducting these amortization charges from the income statement reduces the reported earnings of the company and can depress its share value for as long as forty years.[44]

Unfortunately the amortization charges for goodwill are not tax deductible, they merely reduce the firm's reported income in statements used for financial reporting. So, from the corporation's point of view, there is no advantage to purchase accounting—it merely makes the merger look bad. Under the pooling of interest method, the balance sheet and income statement of the two merging companies are simply added together. If there are merger synergies, the resulting cash flows and financial ratios would indicate a stronger company justifying a higher share price.

The FASB has apparently decided to retain poolings, but also has decided to change the rules to make the purchase method more attractive.[45] To do this, the FASB is presently considering a rule that would shorten the amortization periods for goodwill under the purchase method, or allow companies to not amortize goodwill at all if their acquisitions continue to be worth the premiums paid.[46] As far as international harmonization is concerned, "Why should we use a 40-year [amortization] schedule if many other countries use five?" asked a FASB technical analyst.[47] FASB might also allow companies reporting goodwill to make periodic tests for "impairment." Should market values for companies like the one acquired increase, this would indicate that the amount of goodwill should be reduced.[48] The rule could cut in either direction, though: if the market value of the target company assets decreased, then the goodwill would increase along with the amounts of amortization charges.

MERGERS AND ACQUISITIONS RESPOND TO THE
THE NEW ANTITRUST ENVIRONMENT

Terms Must Produce Lower Rates

The primary directive that FERC has picked up from the new antitrust environment of the Justice Department and the FTC is that mergers must benefit customers rather than utilities. FERC has also encouraged state regulatory officials to take this hard line. The result is that merging utilities are scrutinized to see if the combination provides enough synergies for them to lower their customers' power rates. This section illustrates the response of merger activities to both the liberalized antitrust environment and to FERC's constraints.

Enron Pays Less for Portland

The trend-setting merger of Enron and Portland General Corporation shows how seriously Oregon state regulators took the directive that mergers were to be made for the benefit of its customers. While Enron and Portland were in final merger negotiations, the Oregon Public Utility Commission (OPUC) demanded that Enron and Portland more than double their proposed rate cuts if they were to approve the $2.1 billion merger.[49] To please OPUC, the merger partners increased their rate cuts from $61 million to $141 million. This caused the 1-to-1 rate of exchange of Enron stock for that of Portland General to be reduced to 0.9825-for-1, in effect renegotiating Enron's purchase price for Portland, with Enron providing $81 million of the financing it had originally intended for Portland shareholders instead to the utility customers of Portland as a rate cut.[50] This reduced the value of the proposed merger from $2.1 billion to 2.06 billion.

Divestments and Auctions

Another form of merger and acquisitions activity that deregulation has stirred up is divestments and acquisitions of the generating assets of various utility companies. These divestitures and acquisitions involve the sale and purchase of the power plant's assets that make up the large utilities, rather than the sale or acquisition of the company itself.

PG&E Acquires and Divests Plants

In February 1996, CalEnergy, Duke Power, and the Southern Company considered making bids on all or part of a portfolio of eighteen New England generating plants with a book value of $1.1 billion to be auctioned off by New England Power and Electric Company.[51] Frustrated by aging facilities, including high-cost nuclear plants, New England had decided to become a regional transmission and distribution company instead of generating electricity.

In a surprise acquisition, in August 1996, Pacific Gas & Electric Company (PG&E), the nation's largest power company, gobbled up the assets of New England for $1.59 billion. PG&E's purchase included all eighteen of New England's fossil fuel and hydroelectric power plants, making the first major sale of utility company generating operations under the new deregulation policies.[52] The purchase of these assets by PG&E, itself a low-cost electricity producer, allowed the larger company to enter the rich New England utilities market as New England Power & Electric's wholesale power supplier.

Although PG&E had acquired New England generating assets in August, the following October, it was required to divest power plants in its home state of California. The California Public Utilities Commission (CPUC) required PG&E to divest four of their gas-fired electrical power generation plants for about $400 million.[53] PG&E was one of the three large power producers that had divided up most of the power business in California, one of the richest power markets in the United States. The divestiture allowed the purchaser(s) of the power plants to enter the California market to provide more low-cost, gas-fired competition under the CPUC's plan for deregulating the power supply market in their state.

The California Strategy Laboratory

With the highest electric power rates in the United States, the state of California has been a leader in deregulating the industry suppliers. The CPUC has asked the state's three major utilities to divest themselves of half of their electricity generation plants.[54] This included the divestiture of PG&E's four power plants. Taken in total perspective, California's deregulation plans make up a strategy laboratory for power deregulation policies.

One of the patterns that is emerging from the California strategy laboratory is the strategy of PG&E to specialize in its local monopoly on regional power transmission, divesting itself of home-state power generation, while, at the same time, acquiring power plants in New England to compete there as an electricity generator. New England Power & Electric, the company that sold PG&E its eighteen New England plants, had itself employed half of this strategy—they had specialized in local power transmission but had not, so far, reached out for competitive assets elsewhere in the United States.

PG&E's strategy takes advantage of FERC's willingness to allow giant utilities to merge with noncontiguous utilities in order to enter other states and compete with local power monopolies. California state law has similar goals, and mandates that some California customers be able to buy power from the cheapest sources, *regardless of where the provider is*, by the beginning of 1998.[55] This policy is designed to prevent California's largest power producers from dominating the state pool of power suppliers by creating alternative, out-of-state sources for gas and electricity.

California's deregulation policy began unfolding in January 1997, when regulators unveiled an alliance of eleven cities in California that were to be supplied with electricity by Houston-based Enron.[56] In this alliance, the Northern California Power Agency (NCPA) which supplies power to 700,000 customers, agreed with Enron to buy power from them if it could buy it more cheaply than it could from PG&E. The arrangement was encouraged by the CPUC in order to provide a competitor for PG&E to produce lower power rates. Enron, on the other hand, benefited because it extended its national wholesale power marketing strategy by securing customers in northern California without having to compete with a local utility to get them.[57] This was the same strategy that MCI and Sprint used after the breakup of AT&T. The California market provided a fine test for Enron's strategy since California rates are 50 percent higher than the national average.[58]

Since 25 percent of the electricity in the United States is provided by small power authorities like NCPA, Enron's California alliance was considered another trend-setting deal.[59] It is a new idea of the California strategy laboratory to pit an out-of-state power supplier against an old local monopolist (like PG&E) by matching up the out-of-state company (Enron) with a local monopoly (NCPA) supplying power to various municipalities.

Enron supported its California strategy by unveiling a $200 million a year program to sell its corporate image and the idea of cheaper energy in a deregulated world.[60] This marketing image is expected to make their entry into other states easier. Enron is the first major utility to begin a major advertising campaign of this kind, and it is interesting to speculate on how much consumer preference based on image advertising will matter if Enron is primarily going after wholesale power deals and large industrial customers like those in New York.

Breaking Up Local Monopolies

Despite the modern attitude that allows mergers if they are efficient and have adequate competition—bigness-is-not-necessarily-bad—some states are breaking up their former utility monopolies into competitive units that become specialists in their smaller areas of endeavor. One such case is New York's plan to restructure the huge Con Edison electrical utility. Con Edison is the first of seven of New York's regional utilities to be reorganized and restructured, and the deal is intended to test the pattern for restructuring the other state utility monopolies.

The New York State Public Service Commission and Con Edison management reached a tentative agreement in 1997 to break up the utility into generating, retailing, and distributing companies under one holding company.[61] Separating the generating functions from the more-or-less monopolistic distributing function—the utility's transmission wires—was expected to reduce power rates for industrial customers by as much as 25 percent. These lower rates were expected to stimulate the New York industrial economy and allow the utility to compete with new out-of-state utility giants like Enron.

Strategies of Market Dominance

The current round of utilities mergers has resulted in an unprecedented number of large utility mergers in the United States, so many, in fact, that U.S. mergers have spilled over into the U.K. and Australia. One merger illustrating that big is not necessarily bad is PG&E's 1997 purchase of Valero Energy's natural gas business for $722.5 million in stock.[62] Even though PG&E was already America's largest power company, the merger was approved by FERC. Nor was this PG&E's only recent acquisition. In November 1996, PG&E purchased the Teco Pipeline Company from TRT Holding in Corpus Christ, and in December 1996, they acquired natural gas marketer Energy Source Incorporated of Houston. After the Valero acquisition, Robert D. Glynn Jr., PG&E's president and chief operating officer, said that PG&E now has "a very broad marketing reach that is literally coast to coast and up into Canada."[63]

These mergers and acquisitions were allowed because gas and gas-fired electricity generation are the low-cost means to provide about half of the power in the United States. By retailing this low cost power nationwide, PG&E must engage in competition with merging local power companies. FERC feels that because PG&E is more a more competitive lower-cost producer, in this case, the bigger utility is better.

Bootstrapping Earnings per Share

At least part of the merger strategy of all of the large, low-cost, rapidly merging national power marketing utilities, such as Enron and PG&E, is the financial concept of bootstrapping earnings per share.[64] Bootstrapping is a merger strategy that appears at first to let the merging firms increase their earnings per share merely by merging. This situation occurs where the acquiring company has a price earnings ratio (PE ratio) greater than the PE ratio of the bought company. In many of these situations the acquiring company can offer their shares for the shares of the bought company at an exchange ratio that will offer the bought company stock that is worth more than the shares that they are giving up. In these situations, the earnings per share of the merged company will be seen to increase in the absence of synergies or any other merger advantages other than the arithmetic of the exchange ratios. This is, of course, an illusory increase in value, and it only comes about *if the PE ratio of the merged company is as great as that of acquiring company.*

Although the phenomenon of bootstrapping is always part of the arithmetic involved in calculating the results of a pooling merger, it is unlikely that a utility that repeatedly merged would achieve repeatedly bootstrapped earnings, unless synergies were available to preserve the acquiring company's original, high, PE ratio. But the possibility of this illusion does exist. The illusion would be more likely to occur in the case of a company that merged often, merely because there would be more instances where this kind of arithmetic might occur, unless the market for control were efficient and vigilant in its discrimination between reality and illusion.

Reconfiguring Contracts

Mergers and acquisitions are not the only route to bigger and more efficient power production. Contrast the strategy of AES Corporation (AES) in New Jersey with that of PG&E. AES agreed to buy out the twenty-five-year contracts of three smaller producers to supply electricity to GPU Corporation (GPU) of Parsippany, New Jersey with enough power for 2 million customers.[65] By purchasing GPU's wholesale market for power, AES acquired a market large enough to justify constructing a $300 million 720 megawatt gas-fired power plant in southeastern Pennsylvania. This power plant will produce electricity at a cost of only 3 cents per kilowatt hour as opposed to the 6 to 9 cents per kilowatt hour that GPU had agreed to pay the smaller generators under their contracts.

This arrangement allowed AES to continue the expansion of its large international portfolio of electricity generating assets. AES currently owns thirty-three plants in the United States, Argentina, Brazil, Pakistan, Hungary, Kzakstan, China, and Great Britain and is known as a worldwide developer and operator of power plants.[66] The deal between AES and GPU repeated a pattern of utility development that had proven successful for AES in the past. Typically, AES would put together the financing for its power plants which were then built by outside contractors and operated by themselves.

FERC approved this deal, although it consolidated the electrical power supply for three regional areas in New Jersey and Pennsylvania, because the AES contract halved the wholesale cost of GPU's electricity and would allow them to reduce their rates for approximately 2 million customers. Although the three regions may have been contiguous, GPU was already supplying them, so the new contractual arrangement did not make the situation either less competitive or more monopolistic.

Although GPU owned three utilities—Jersey Central Power and Light Company, Metropolitan Edison Company, and Pennsylvania Electric Company—it chose to contract out electricity at rates that it felt that it could not, itself, produce. The new contractual arrangement with AES allowed GPU to escape long-term contracts with smaller power producers that were 100 percent above market.[67] This contracting arrangement is similar to Oglethorpe Power and Light's deal to buy over half of their electricity at low cost for at least fifteen years from Louisville Gas and Electric Company (LG&E). From a financial point of view, GPU's contract with AES amounted to a large, long-term call option on the future price of electricity. For AES, which sold the call, the contract behaves like a put on the future price of electricity, with AES betting that the price will go lower than the 3 cents per kilowatt hour stipulated in its contract.

The contractual arrangements between GPU and AES, and Oglethorpe and LG&E, show how utilities trade long-term energy contracts instead of engaging in corporate mergers. Instead of acquiring GPU, AES, has *sold* them a twenty-five-year call on electricity. GPU, running older, less efficient electric plants has *bought* the call, betting that electricity will rise above 3 cents a kilowatt hour, allowing them to buy cheap and sell dear to customers at if prices go up to, say, 6 cents per kilowatt hour. Notice that GPU will make a 100 percent profit on this contract if the price goes to 6 cents per kilowatt hour, and the gamble involved no change in managerial control for either party.

Local Mergers

Wisconsin Energy Corporation dropped its plans to merge with the Northern States Power company when FERC asked them to sell some of their plants to gain approval of the merger.[68] The request for divestiture was made of Wisconsin under FERC's new guidelines for avoiding too much market concentration.

The circumstances of the Wisconsin-Northern States merger were immediately compared to the large merger pending between Louisville Gas & Electric Company (LG&E) and Kentucky Utilities Company (KU).[69] LG&E argued that the merger that they proposed was a much smaller, $1.43 billion, than the Wisconsin-Northern States combination. They also stressed that they had achieved their status as of one of the lowest cost national marketers of electricity based on their abundant sources of cheap coal and efficient coal fired generation. Also in its favor was its local market situation, for although LG&E was a national marketer of both electricity and gas, it could claim only a modest home market and was surrounded by other powerful competitors. FERC had approved similar mergers of other large utilities, like Enron and PG&E which had aspired, like LG&E, to become national low-cost power marketers.

On the other hand, if LG&E acquired its rival, KU, the merger would produce a company with revenues of $4.3 billion and 1.1 million customers.[70] The merger would also produce a moderate degree of local market power in an area served by the two utilities. What FERC is faced with in the case of LG&E and KU is a merger that might not have been approved under the big-is-bad rules, but which is currently balanced in favor of approval because of the additional national competition that the growing power of LG&E's low cost generation provides. FERC will probably allow the merger and catapult LG&E and it's partner, KU, squarely into the national competitive arena. Under the current antitrust environment a negative decision would only brand LG&E as an overly ambitious local monopolist, and this is not the way that deregulators treat low cost utilities today.

Merger Solutions for Stranded Costs:
Merging the Strong with the Weak?

Unregulated price competition in energy means that utilities with costs stranded in their formerly protected, high-cost production methods—such as nuclear power plants—will become increasingly risky. The merging utilities are bargaining vigorously with power regulators, state legislators, and FERC to see who will bear the cost of refinancing huge amounts of capital investment sunk in often unusable and always expensive nuclear power plants.

Unlike the banking industry, where the Federal Reserve Bank encouraged strong banks to merge with and take over failing savings and loan associations, the utilities industry, with its free-wheeling value-conscious mergers, is unlikely to encourage large low-cost power producers to saddle themselves with the stranded costs of weaker firms' nuclear production assets.

The low-cost power marketing utilities will certainly not merge with weak utilities on their own initiative. Strong utilities are doing exactly the opposite; they are seeking out the lowest cost, highest profit merger partners that they can find, both domestically and internationally. In 1996 and 1997, they have done this with such intensity that the U.S. market is beginning to look picked over. It seems more likely that the stockholders and bondholders of weak utilities will bear the cost of their utilities' stranded assets in the form of depressed share values, increased bond risk, and bankruptcies unless local governments can somehow refinance the costs and spread them out enough to let the weak producers compete with the national power marketing utilities.

Having said this, it is only fair to point out that the merger between the Long Island Lighting Company (LILCO) and the Brooklyn Union Gas Company is a merger of a strong gas company with a weak electrical generator in one of the most high profile cases in the United States.[71] The only factor mitigating this merger is that the merger plans include a provision to refinance LILCO's stranded costs.[72]

Who Bears the Burden?

There will be continued negotiations between ratepayers and the stock and bondholders of utilities to see who bears most of the burden of stranded nuclear costs. In this contest, the rate payers are the likely losers, because—as in the LILCO-Brooklyn Union merger—they are less well organized than the utilities investors. Another reason that most of the stranded costs will be paid by utilities customers is because the regulatory authorities perceive the need to support utility share and bond prices to keep the utility's securities market-worthy enough to finance them out of the hole that the stranded nuclear costs have put them in.

If regulators force the burden of refinancing onto rate payers by sustaining power rates above market prices, in a competitive market, this will decrease cash flows to the utilities and their share prices will fall and bondholders will see their utility bonds downgraded. If stranded power plants are refinanced with tax-free bonds, this eases the risk of the bondholders at the expense of statewide tax-payers who might never have been served by the obsolete plants. But as the re-financing of these obsolete assets is resolved, electric power users will see moderating pressures for rate increases, and in the longer term, real reductions in rates.

NOTES

1. Catherine Yang, "Regulators: This Top Trustbuster May Tone It Down," *Business Week*, October 7, 1996, p. 36.

2. *Ibid.*

3. *Ibid.*

4. Peter Passell, "A Sea Change in Policy by the Trustbusters," *New York Times*, March 20, 1997, pp. D1, D2.

5. *Ibid.*

6. *Ibid.*

7. *Ibid.*

8. *Ibid.*

9. Charles H. Studness, "Converging Markets: The First Real Electric/Gas Merger," *Public Utilities Fortnightly*, October 1, 1996, 134(18): pp. 21-25.

10. *Ibid.*

11. *Ibid.*

12. John R. Wilke, "FTC Says Staples' Bid for Office Depot Sought to Remove Most Aggressive Rival," *Wall Street Journal*, May 20, 1997, pp. A1, C21.

13. Passell, "A Sea Change," pp. D1, D2.

14. James P. Miller and Benjamin A. Holden, "Northern States Power and Wisconsin Energy Called Off Their Merger, Citing the Unexpected Regulatory Complications They've Encountered," *Wall Street Journal*, May 19, 1997, pp. A1, A4; and James P. Miller and Benjamin A. Holden, "Wisconsin Energy-Northern States Pact Rejected," *Wall Street Journal*, May 15, 1997, p. A1.

15. Passell, "A Sea Change," pp. D1, D2.

16. *Ibid.*

17. *Ibid.*

18. *Ibid.*

19. *Ibid.*

20. John R. Wilke, "New Antitrust Rules May Ease Path to Mergers," *Wall Street Journal*, April 9, 1997, pp. A1, A3.

21. *Ibid.*

22. *Ibid.*

23. *Ibid.*

24. *Ibid.*

25. *Ibid.*

26. *Ibid.*

27. John R. Wilke and Bryan Gruley, "Merger Monitors: Acquisitions Can Mean Long-Lasting Scrutiny by Antitrust Agencies," *Wall Street Journal*, March 4, 1997, pp. A1, A10.

28. *Ibid.*

29. *Ibid.*

30. *Ibid.*

31. *Ibid.*

32. *Ibid.*

33. *Ibid.*

34. *Ibid.*

35. *Ibid.*

36. *Ibid.*

37. *Ibid.*

38. Securities Data Company, in Elizabeth MacDonald, "Merger Accounting Method Under Fire," *Wall Street Journal*, April 15, 1997, p. A4.

39. MacDonald, "Merger Accounting Method," p. A4.

40. Elizabeth MacDonald, "FASB May Back Off From Its Threat to Limit or End 'Poolings of Interest,'" *Wall Street Journal*, July 1, 1997, pp. A1, A4.

41. MacDonald, "Merger Accounting Method," p. A4.

42. *Ibid.*

43. *Ibid.*

44. *Ibid.*

45. MacDonald, "FASB May Back Off," pp. A1, A4.

46. *Ibid.*

47. MacDonald, "Merger Accounting Method," p. A4.

48. MacDonald, "FASB May Back Off," pp. A1, A4.

49. Terzah Ewing, "Enron and Portland General Reduce Stock-Swap Ration and Boost Rate Cuts," *Wall Street Journal*, April 15,1997, p. A6.

50. *Ibid.*

51. Ross Kerber, "Auction of 18 Power Plants is Igniting Utilities' Interest," *Wall Street Journal*, February 6, 1997, pp. A1, B4.

52. Charles V. Bagli, "PG&E Will Buy 18 Power Plants in New England," *New York Times*, August 7, 1997, pp. D1, D2.

53. Agis Salpukas, "Pacific Gas and Electric to Sell 4 Power Plants," *New York Times*, October 23, 1996, p. D4.

54. *Ibid.*

55. *Ibid.*

56. Peter Fritsch, "Enron to Unveil Energy Alliance Inolving 11 Cities in California," *Wall Street Journal*, January 15, 1997, pp. A1, B4.

57. *Ibid.*

58. *Ibid.*

59. *Ibid.*

60. *Ibid.*

61. Richard Perez-Pena, "Pact Reached to Break Up Con Edison," *New York Times*, March 14, 1997, pp. B1, B4.

62. Benjamin A. Holden, "PG&E Agrees to Buy Unit from Valero," *Wall Street Journal*, February 3, 1997, pp. A1, A3

63. *Ibid.*

64. James C. Van Horne, *Financial Management and Policy*, 11th ed. (Upper Saddle River, N.J.: Prentice Hall, 1997), pp. 597-598.

65. Agis Salpukas, "Utility Deal Aims to Cut Cost of Power," *New York Times*, February 5, 1997, D1, D5; and Holden, "PG&E Agrees," pp. A1, A3.

66. *Ibid.*

67. Agis Salpukas, "Utility Deal Aims to Cut Cost of Power," *New York Times*, February 5, 1997, pp. D1, D5.

68. Allen R. Myerson, "LG&E Energy Agrees to Buy Rival Utility in Kentucky," *New York Times*, May 22, 1977, pp. D1, D21.

69. *Ibid.*

70. Benjamin A. Holden, "LG&E to Buy KU for $1.43 Billion in Stock," *Wall Street Journal*, May 22, 1997,pp. A6.

71. Bruce Lambert, "Stockholders of Lilco and Brooklyn Union Ratify Merger of Companies," *New York Times*, August 8, 1997, p. B15.

72. *Ibid.*

Chapter 5

Merging Natural Gas and Electricity to Produce Power More Effectively

NATURAL GAS AND ELECTRICITY AS SUBSTITUTES

In the United States, electrical power companies have been merging with gas companies in order to market their power more effectively. Large industrial customers now have the ability to use electricity and gas as substitute power sources, an ability that allows them to switch to whichever source has the lowest price. The energy utilities also are able to switch the raw materials they use to produce electricity in the most efficient way. This means that, increasingly, U.S. utilities find themselves switching to gas-fired electricity generation because it is becoming cheaper than coal-fired production. The economics and technology of power production are making electricity companies and gas utilities a natural combination.

The increased substitution of both energy sources and the inputs used to generate energy has been termed *energy convergence*. Coal-fired plants convert coal into electricity, producing "solid electricity." Gas-fired plants generate electricity by burning regasified liquid natural gas (LNG), which produces *gaseous electricity*. Utilities swap their expensive electricity supply contracts for cheaper coal production contracts, or vice versa, depending on the market prices of coal and electricity.[1] The companies that trade these contracts may operate in different countries, each trading their energy product, which is produced locally for a low price, for a foreign country product that could be produced locally only at a premium.[2]

Gas-to-Electricity Mergers

The first large merger in the United States to be inspired by the concept of gas-to-electricity convergence was the $2.1 billion acquisition of Oregon's Portland General Corporation in July 1996 by Texas-based Enron Corporation.[3] The trend was continued when Duke Power and PanEnergy entered a $7.7 billion merger the following November.[4] When CalEnergy acquired U.K.-based Northern Electric PLC in October of 1996 for $1.55 billion, they took the trend international.[5]

Coal-to-Electricity Mergers

Probably the most interesting, if not the largest, energy convergence merger was PacifiCorp's $6 billion pending bid for Energy Group PLC in the United Kingdom. Begun in June 1997, this deal is interesting because it is an example of an international merger involving coal-to-electricity convergence where the U.S. corporation appears to be buying not only cheap coal production, but also the ability to arbitrage both coal and electricity contracts internationally.[6]

The New Technology of
Gas-to-Oil Conversion

The new Fischer-Tropsch gas to oil conversion processes illustrates how new technology is making possible economical conversions between almost all major energy sources. Gas was previously flared at the well head because, compared to the petroleum beneath it, gas was not a profitable product. The Fischer-Tropsch process allows burning gas into oil at the well head by using a catalyst to convert the gas into oil using the heat from the combustion of the gas.[7]

The process was clumsy at first, but Exxon Corporation has quietly invested over $100 million to improve the catalysts used in the Fischer-Tropsch gas-to-oil conversion process, bringing down the cost of producing a barrel of synthesized oil from $35 to $15-$20 a barrel.[8] Kenneth L. Agee, working in the Tulsa laboratory of Syntroleum Corporation has developed a proprietary catalyst:

> Agee adds a little magical "eye of newt"—a proprietary catalyst that spins out only short-chain hydrocarbons such as naphtha and kerosene. Avoiding the wax "pudding" means the light crude will readily flow in oil pipelines. Although the light products are less valuable, the process could be just the ticket for sites such as Alaska's North Slope, where pipelines already exist. Such plants may cost as little as $14,000 per barrel of daily production, or half the investment for supercooling the gas into liquefied natural gas (LNG).[9]

Tiny Syntroleum has licensed this process to Arco, Texaco, and USX's Marathon Oil.

The Sasol Oil Company began to use the Fischer-Tropsch chemistry in Johannesburg to convert gasified coal to oil after petroleum shipments to South Africa were embargoed by the West in the mid-1980s. Since then, the new catalysts have allowed gas producers to eliminate the expensive cryogenic machines that formerly removed oxygen from the gas by substituting a one-way ceramic membrane. Besides reducing the production costs (by eliminating the oxygen separation step) the new techniques cut the cost of capital equipment by 25 percent. Once the gas has been converted into oil, it is not only cheaper to transport, but also has a ready mass market. [10]

THE EMERGENCE OF NATURAL GAS

Worldwide gas production is increasing twice as fast as the rate of oil production and faster than consumer demand. Russia and the Mideast are far ahead of other producers in gas production. The Mideast, although it produces only one-fifteenth of the world's gas, has fully one-third of the known natural gas reserves—mostly in Iran, Saudi Arabia, and Qatar. Nearly $40 billion of export projects in gas production are either planned or under construction here, and they are expected to cause a gas boom much like the oil boom of the 1960s.[11] But this boom is quite different from the oil boom. "Gas is a whole new ball game," says Julia Nanay of Petroleum Finance Co., a Washington energy-consulting firm. "It makes a country and customer dependent on each other in a way that oil doesn't."[12]

Qatar's Huge Head Start

Qatar is interested in becoming the richest nation in the world on the basis of its new gas field in Ras Laffan, an area that had no paved roads until recently. Talk of rapid development makes U.S. officials nervous about friction between Qatar and its other Mideast neighbors, especially since Saddam Hussein's invasion of Kuwait on the pretense of oil field poaching.[13]

Adding to diplomatic worries is the fact that Iran and Qatar compete for the same gas. Iran and Qatar have plans to tap the same gas field in South Pars. If Iran does not develop the gas, Qatar will tap into it. Because gas flows more freely underground than oil does, "it's physically impossible to divide it" while it is in the ground, says an executive in one of Qatar's gas projects. "Once both countries insist on ownership, it becomes a political problem."[14] So far, Iran is merely observing Qatar's gas drilling rigs from their navy vessels.

It is crucial to liquefy the natural gas, because shipping gas is expensive and it is much cheaper to ship a compact liquid than a gas.[15] Natural gas is compressed by cooling it to 160 degrees below zero centigrade, at which point it turns into a liquid—liquid natural gas. Liquid natural gas can then be loaded onto specially insulated ships, and "regasified" in the receiving country. For trips of less than 1,000 miles, pipelines are cheaper, but even so, gas pipelines are more expensive than oil pipelines and they must sometimes pass through hostile or unstable countries.

Currently the supply is greater than the demand for natural gas. Qatar is charging only a token royalty of $60 million a year for the gas in each "train," or array of 1,000-yard gray pipes and towers that liquefy the gas.[16] There are not

even enough customers to create a world standard price for Mideast gas. Qatar's
Ras Laffan Liquefied Natural Gas Co. (RasGas) first agreed with South Korea
that the price for LNG would not fall below $14.44 a barrel, the price of of Qa-
tari crude oil. This let RasGas raise $1.2 billion through bond offerings to devel-
op their gas plants. But RasGas later had to renegotiate the contract with no price
floor after South Korea made a deal to buy Oman's gas with no floor. The contract was renegotiated in return for an agreement to buy twice as much gas. Mobil, on the other hand, promised to pay as much as $200 million to the bond-
holders if oil prices fell so low that RasGas could not meet its bond payments.

Economies of scale for natural gas are huge and high transportation costs are
high. Despite the high-cost of development and soft gas prices, countries are
competing for both financing and for customers that can consume the output of
the multibillion-dollar gas projects. As a result, Russian and Mideast natural gas
producers are undercutting each other's prices in favor of larger volumes, just as
oil producers undercut each other by exceeding production quotas.

Qatar is diversifying into the production of natural gas. "In the '80s, we learned our lesson, not to put all our eggs in one basket," says Abdulla Al-Attiyah,
Qatar's minister of energy and industry.[17]

Fluctuating Prices and Long-Term
Contracts—Risk for Companies

Typical contracts for natural gas production run for as long as twenty-five
years. With terms like these, the likelihood that prices will change makes private
gas projects very risky. If a large buyer defaults, Mideast gas can not easily be
sold elsewhere. So as gas becomes a more important global fuel the goal is to
monetize the gas by turning it into money. As soon as global markets develop to
the point where there is a Mideast price for gas, floating price contracts can be
made (much like eurocurrency loans) by letting the contract price float above the
Mid-east price (much as a Euroloan's rate floats above the London interbank
offer rate). The Mideast market will develop because, says Fritz Volgt, vice
president of the international gas unit of Exxon Corporation, "No one can afford
to ignore gas anymore."[18] Exxon has more gas than oil in its undeveloped but
proven reserves.[19]

An example of the kind of risk that global utilities are facing in terms of gas-
to-electricity conversions is Enron's $440 million negotiated settlement to back
out of a North Sea gas deal. Enron had agreed to pay Phillips and its partners,
British Gas PLC and Italian Agip SPA, more than $3 dollars per thousand cubic
feet of gas.[20] Enron hoped to sell gas from a North Shore production area known
as the J-Block to new U.K. power plants. To do this, Enron entered the 1993
contract with Phillips, intending to buy about 800 billion cubic feet of gas. Then
natural gas prices fell. Enron claimed that the resulting $675 million charge
would equal a loss of about $1.80 a share, that the new price for the gas would
be "significantly less" than the originally contracted price.[21]

Gas Investors

There is an impressive array of countries beginning to develop huge new natural gas resources. The following is a brief synopsis of how some of these countries are participating in the opening of massive new natural gas resources.

Qatar

Qatar owns $1 billion worth of majority stakes in the gas companies of the Ras Laffan gas field. The companies themselves have invested another $10 billion in their operations. Qatar also owns the RasGas.[22]

Indonesia

In September 1997, the Atlantic Richfield Company (ARCO), announced a $3 billion investment to develop the Wiriagar gas field in Indonesia's easternmost province.[23] The field was estimated to contain 13 trillion cubic feet of gas, and was ARCO's largest field outside of the United States. ARCO's investment in Wiriagar would include the development of the Tangguh liquification plant, which was expected to produce 6 million tons of liquefied natural gas by the year 2003. This investment would allow ARCO to compete with Mobil and Total SA of France, which now dominate the liquefied natural gas business in Asia.

United States

Exxon has invested over $100 million in its vastly improved Fischer-Tropsch process to convert gas into oil. Exxon has focused this gas-to-oil conversion process on plants that can produce up to 100,000 barrels of synthetic crude daily.[24] There are only half a dozen places where the gas fields are sufficiently large to support this scale of production. Of these, Russia, Iran, Saudi Arabia, and Qatar and the most likely to be develop. Since Russia has been preempted by Gazprom, the huge Russian state-owned producion company, and Iran is off limits because of U.S. policy, the only remaining qualified gas fields are those of Saudia Arabia and Qatar. In October 1996, Exxon announced that it was negotiating with Qatar General Petroleum Corporation (QGPC) to build a $1.2 billion conversion refinery with an initial capacity on the order of 50,000 barrels a day.[25] "If you weren't paying attention [to conversion technology] before Exxon's announcement, you were then," said Ralph A. Avellanet, manager of Energy's gas-processing program.[26]

France

Total SA of France has formed a partnership withFrance Mobil Corporation of Fair-fax, Virginia, to create Qatar's first gas gas export company. But Total got left out of the second phase of development at Ras Laffan, when Qatar moved closer to the United States—and Mobil—after the Gulf War. This situation is reversed in Iran, where Mobil is locked out of Iranian gas production, because of U.S. sanctions preventing Americans from doing business there.[27]

Malaysia

Malaysia's Petronas oil company may join the team of Total SA (France) in their gas development projects.

Britain/Netherlands

Royal Dutch Shell discovered the South Pars field in 1971 while searching for oil. Now Anglo-Dutch (Britain-Netherlands), Royal Dutch/Shell Group (Britain-Netherlands), and Total SA (France) are all striving to develop the field.

Gas Politics

U.S. Sanctions on Iran

The U.S. sanctions on Iran are a major factor complicating the development of Mideast gas fields. France's Total SA ignored the sanctions and proceeded with its investments in Iran's development of the South Pars field, as have Gazprom, Petronas and the Malaysian Oil Company, which may join Petronas' team.[28] Shell, with extensive U.S. operations, has taken no *public* position on Iranian gas development, although it has held discussions with the Iranians. Qatar, which is developing into a U.S. military ally, has maintained friendly relations with Iran despite the fact that they are Iran's major gas competitor with common borders on developing gas fields like that of South Pars.[29] International banks and credit agencies are under strong U.S. pressure to stay out of Iran.[30]

Currently the main foreign investors in Iran's South Pars gas field are Anglo-Dutch Royal and the Dutch/Shell Group. This consortium has developed the location of the richest gas deposits along the Iran-Qatar border.[31]

Border Conflicts

The development of the Ras Laffan and South Pars gas fields by Iran and Qatar are governed, if they are governed at all, by a twenty-eight-year-old agreement on where the border of Iran and Qatar lies.[32] Even if the two countries could agree on the position of the border, such an agreement wouldn't determine how much gas could be drawn by each side. Slant drilling technology is very sophisticated today, and the gas flows into underground crevices more easily than petroleum does, making the access points to gas fields more numerous.

The prospectus for the RasGas bond offering shows only a small sliver of the field on Iran's side of the border, but Iran plans to draw 300 trillion cubic feet or more from the field. Also, though the field is huge, the best gas lies in a section called "K-4," straddling the border, and Qatar's RasGas started drilling there first, less than 10 miles from the border. "I'm sure 300 [trillion] is the minimum," says S. E. Jalilian, director of offshore development for the National Iranian Oil Co. "Based on our new calculations, it should have increased."[33] Perhaps these problems are one reason that Egypt has not agreed to attend this year's Middle East Economic Summit in Qatar.[34]

Pipelines

Russia

Russia would like to be the premier gas supplier to Europe, and to that end plans to build the multibillion-dollar Yamal pipeline to Europe. Russia's gas monopoly, Gazprom, controls virtually all Russia's natural gas pipelines, and Russia and Iran appear destined to be competitors for the European gas market as Iran plans a pipeline into Turkey, a market that is presently dominated by Russian gas. Since Russia would prefer that Iran's gas flow eastward, Gazprom recently offered to help Tehran build an liquefied natural gas plant to ship offshore gas to Pakistan, India and China.

Iran

Although the U.S. has tried to talk Turkey out of approving the pipeline, construction has already begun. Iran is trying to get Pakistan and India to agree on a pipeline too, although it considers both of these countries to be adversaries. Iran's new pipeline plans an incursion into the Russian market for gas in Turkey. Pakistan's military authorities do not want Iran surveying their coastline, and India does not want the pipeline to follow an overland route controlled by Pakistan.[35]

Dubbed "Project Peace," Iran plans to cement relations with Armenia by piping its gas there too, despite Armenia's lack of hard currency.[36] Iran is also helping Turkmenistan to build a gas pipeline to Turkey. Perhaps because Iran fabricates its own gas pipe, and despite the fact that Iran is building its own pipeline to Turkey, Iran would like to complete this pipeline by the end of 1997 to get Turkmenistani gas to Turkey.[37]

Unocal Corporation of El Segundo, California, plans to build a gas pipeline from Turkmenistan to Afghanistan and Pakistan that would bring Turkmenistan's gas through these countries to Turkey,[38] which would compete directly with the pipeline being built by Iran.

Financing a Russian Gas Giant

A recent western bond offering by Russia's natural gas monopoly, Gazprom, gives some idea of the power of Russian gas production. In May 1997, Russia announced that it was seeking to raise $2 to $3 billion in Western capital to finance the projects of RAO Gazprom, the Russian gas monopoly that controls 95 percent of Russia's natural gas pipelines and production.[39] From a global perspective, Gazprom controls one-third of the world's natural gas reserves and has revenues of $20 billion.[40] One of the largest projects on Gazprom's agenda is the multibillion-dollar Yamal gas pipeline to Europe. During the next six months that followed, the Russian government attempted to convince foreign investors to buy shares in Gazprom, in which the government owns a 40 percent share.

Gazprom's eurobond offerings, made primarily in western hard currencies rather than in rubles, could be worth more than the Russian Federation's own recent debut debt offering. Goldman, Sachs & Company and the Dutch bank ABN-

Amro planned to issue at least $1 billion in Eurobonds and another $500 million to $1 billion in convertible bonds to finance Gazprom's project.[41]

Although Gazprom's securities were expected to sell briskly, only recently have Russian stocks been considered investment worthy in the west. Things turned around in 1996, when when Russian companies began issuing American Depositary Receipts (ADRs), and in 1997, the Moscow stock market went up 124 percent.[42] The securities of Russia's Gazprom made the second spot on the *Business Week* list of the 100 best emerging market companies, behind number one Telebras, the giant Brazilian telephone company.[43] The boom in Russian stocks has been driven partly by a boom in foreign funds, and partly by the July 1996 reelection of Boris Yeltsin. After his election, Yeltsin reassured foreign investors of their Russian property rights. In response, foreigners pumped almost $3 billion into Russian stocks in 1996 and added another $1 billion in 1997.[44]

But the appearance of Gazprom on the *Business Week* list is partly due to the sheer size of the Russian company. "We're not talking about bread makers here. These are mammoth industrial enterprises."[45] The new Russian stocks on the *Business Week* scorecard included some of the most transparent companies on the market. Lukoil, for example, recently published western-audited figures for 1994 and 1995 and planned to release its 1996 numbers in July of 1997. Even secretive Gazprom planned to release western-quality financial reports. This behavior is transforming both companies into western-style global players.[46] Gazprom has been so successful that astute Russian investors are looking beyond Gazprom to smaller telecom companies and utilities.

THE CHANGING STRUCTURE OF THE GAS INDUSTRY

The global gas industry is rapidly being restructured. There is an impressive list of reasons why this is so:

- The industry is undergoing significant technological change.
- Gas-fired generation is becoming the lowest cost way to produce electricity.
- Energy convergence technologies allow natural gas to be economically converted into liquid natural gas, electricity, and oil.
- The entire industry is being privatized.
- The Russian giant Gazprom is rapidly expanding production at the same time that massive new Mideast gas fields are being opened.
- Modern energy markets are beginning to monetize gas reserves.
- Oil and gas companies are attempting to apply the lessons they learned from the last gas bubble, in order to avoid overdeveloping the emerging gas fields.
- The utilities industry, which consumes much of the natural gas is changing:
    - electrical utilities that use natural gas are being privatized;
    - electrical utilities that use natural gas are being deregulated; and
    - natural gas production is being deregulated.
- As electrical utilities become global, gas producers become more dependent on their production companies, markets, and customers, presenting a scenario of globalization-integration-dependency.

New technology is making possible both lower-cost gas-fired electricity generation and the conversion of natural gas into oil. Both oil and low-cost gas-fired electricity have huge global markets. The provision of natural gas to the mass markets of electrical utilities and automobile drivers is bringing on-line massive new gas fields in Russia, Iran, Saudi Arabia, and Qatar. The global deregulation of electricity generation and the simultaneous privatization of the utilities industry is making it possible for efficient markets to allocate energy optimally, world-wide, for the first time. Utilities and the producers of oil and natural gas are engaging in increased amounts of world-wide arbitrage in both the inputs and the outputs of energy production. Global utilities and gas and oil developers are attempting to buy low and sell high in such a way that the cheapest input is used to generate electricity, and, as a result, competitive markets are beginning to guarantee that the consumer gets the best possible electricity rates given the scarcity of the world's energy resources.

Despite the continued use of long-term energy contracts, the development of global markets in all energy products and raw materials is producing more interdependence between the owners of oil fields, natural gas fields, electrical utilities, nations, and consumers. This must be so because although free markets in energy products produce the maximum freedom possible for deal making, global energy markets and increased energy convergence also mean that decisions on energy projects cannot be made in isolation. Energy producers must estimate much more carefully whether or not the project costs will allow the production of a competitive new product

Privatization

The privatization of state-owned natural gas companies is taking a different approach, because it is coming at the same time that the global market for electricity is being deregulated.[47] Moreover, the privatization of state gas companies in the U.K., Australia, the United States and other countries is based on the economic theory that privately-owned companies operating under conditions of unregulated market prices can simultaneously produce cheaper power and greater profits because of the greater operating efficiencies that competition encourages. It is increasingly evident that as energy mergers continue, the companies that will drive the industry will be more efficient, better managed global utilities. These merging companies already know that market share is becoming more important in the world's increasingly competitive energy markets. So they are restructuring themselves by scouring the globe for low-cost, high-return acquisitions to shore up their competitive positions.

In this new privatized world of global utilities, shareholder needs will come first. Financing is more important now that these companies no longer receive government aid.[48] But financing has always been important. This is because share prices determine the value of the utilities, and the greater the share value of a utility the more capital the utility's new stock offering allocates to them rather than to their competitors. Access to capital is especially important now, not only because it is needed for the acquisitions and strategic alliances utilities find necessary to optimize their global operations, but also to pay for technologically up-to-date generating equipment and advertising programs focused on new mass markets for power. Companies like Repsol, British GM, and YPF, are putting shareholder needs ahead of the goals of their former government owners. These goals include expansion abroad and significant investments in generating power, infrastructure, and additional gas reserves.[49] All of these risky initiatives are taken to increase the utilities' rates of return.

The concurrent restructuring of the gas industry and the probable shift of much of the world's electricity generation to gas-fired plants also creates changes in priorities from those which historically drove utilities. This is true whether the utilities were publicly-owned or not. One of the forces for such restructuring can be seen as both Russia's Gazprom[50] and Mideast gas producers[51] ramp up their production of natural gas. The wider availability of this new low-cost energy input will restructure both the production, pricing, and marketing of global electricity.

Technological Change

Nowhere does technological change impact corporate structure, industrial structure, and market prices[52] more dramatically than the impending conversion of gas-to-LNG, gas-to-oil,[53] and gas-to-electricity.[54] In general terms, this is an example of energy convergence:[55] production technology and the financial technology of energy markets are conspiring to create global market prices for energy equivalents of all kinds. One major change area in energy convergence is that of international utilities merging to acquire the global positions from which they can arbitrage a utility's energy resources. Another is the conversion of natural gas into both oil and electricity. The conversion of gas into LNG is important too, but somewhat old hat, because it has always been necessary to convert natural gas to LNG to keep the transportation component of its cost low enough to make it a competitive energy export.

Very new, however, is the massive conversion of natural gas into both oil and electricity. Both of these conversions dramatically lower the transportation cost involved in selling natural gas. If the conversion of natural gas into LNG lowers its volume and decreases its transportation cost, then its conversion into oil using the Fischer-Tropsch process[56] also reduces its volume and brings down its transportation cost—even more. But when gas is converted into oil, the oil comes with a ready-made global market as well. Since one of the major reasons that natural gas has been ignored is that it has traditionally been a low-profit commodity, mass markets are particularly important to make it economical enough to justify bringing onstream the major new gas fields in Iran, Saudia Arabia, and Qatar.[57] The importance of gas conversion technologies can hardly be overstated.

Perhaps even more interesting than the conversion of gas-to-oil is the conversion of gas-to-electricity. Whereas the conversion of gas-to-oil involves a clever but straightforward application of petroleum technology, the conversion of gas-to-electricity involves the technology of global energy markets.[58] Once the natural gas has been burned in a gas-fired turbine generator, the contract for the resulting electricity can simply be traded, at global market prices, anywhere in the world. The transportation cost for this natural gas, once converted, is often close to zero plus a brokerage fee. The brokerage fee is not applicable today, because these electricity contracts are traded and arbitraged primarily between multinational firms and subsidiaries making up the global utilities industry, but formal energy markets will develop soon. If the utility does not trade energy contracts, its alternative would be to put the electricity on the transmission grid and sell "electrical gas" or "electrical coal," depending on which fuel was burned to generate the electricity.

All of these changes are reducing the price of energy. Generally, the technologies that allow the substitution of one energy source for another will make it more likely that a global utility will be using the lowest cost energy input—coal, oil, gas, or nuclear power—and selling lower priced energy based on this decision.

New technologies also reduce production costs by getting the job done more efficiently. What could be more ingenious than Exxon's substitution of a new ceramic tissue for the old, large, expensive, cryogenic oxygen-separation machinery in the Fischer-Tropsch gas-to-oil conversion? What could be more useful than the fact that this technology will allow Qatar to develop its impressive natural gas resources? The answer, if you have difficulty with finance, is the idea called the *monetization of gas-in-the-ground*. Modern markets, by quoting prices for energy equivalents of gas, will make it possible to compute the net present value of undeveloped proven natural gas reserves. Once their value has been priced, these reserves could be securitized and traded like any other asset. This would allow financial markets to arrange for the financing of new gas fields in much the same way that they would finance a manufacturing firm's investment in machinery.

Lessons Learned from the Gas Bubble

The rapid development of massive new gas fields in Iran and Qatar, and the aggressive expansion of Gazprom into western markets, remind gas experts of the gas glut that developed in the 1980s. This "gas bubble" resulted in low prices that were supposed to be short-lived, but these prices are still with us in 1998.[59] U.S. gas producers waited for gas supplies to be deregulated so they could make a killing. But gas is still regulated and Mesa Petroleum, for example, almost went bankrupt. This time around, Morgan Stanley researchers Sack and Chewning suggest that the longer it takes the energy market to reach an equilibrium (world) price for natural gas, the lower the price of natural gas will fall.[60] The lessons of the gas bubble are cautionary tales about forecasting high energy prices. These tales also involve the suspicion that the development of the Russian and Mideast natural gas fields may result in a condition of oversupply that will bring down the price of natural gas once again.

It is appropriate to remember the 1980s gas bubble now because current world fuel reserves could mushroom by an astronomical figure if the reserves were to be exploited immediately. Worldwide gas reserves have increased almost 6 percent a year since 1970, but consumption has risen only 3.1% annually. This has resulted in a new gas bubble of 41 quadrillion cubic feet of gas in known reserves, according to Geneva-based Petroconsultants.[61] That represents 770 billion barrels of oil equivalent, enough to slake the world's thirst for oil for 29 years.[62]

The Dependency Hypothesis

As electrical utilities become global, gas producers are becoming more dependent on their production companies, markets, and customers. Because of this pattern of relationships, the development of new natural gas production is presenting a scenario of globalization-integration-dependency. Utilities are becoming globalized, energy markets are becoming globally integrated, and gas producers are becoming more dependent on their production companies, markets, and customers.

This dependency is developing because the gas boom is not simply a repeat of the oil boom. One reason for the dependency is that gas is in excess supply, and access to large markets of dependable gas consumers is at a premium.[63] Retaining these customers and markets is particularly crucial while producing countries are still in the process of arranging the financing to open huge new fields, at a time when cash flows have not yet materialized.

Pre-Processed Gas

The financing of natural gas projects is especially crucial because LNG equipment is expensive. But it is absolutely essential to ship natural gas as a liquid in order to keep down its transportation cost at competitive levels. Although pipelines are cheaper for distances of less than 1,000 miles, they are more expensive than oil pipelines and must sometimes pass through hostile or unstable countries. Qatar, for example, cannot build a pipeline to the United States because of the distance, and pipelines from Qatar to other countries are fraught with political problems. In this respect, gas production can be compared to oil production, but it must be refined before it can be shipped.

Excess Supply and Price Competition

Competition in all of these areas makes producers more dependent on their customers than usual. This is especially true in the pricing of gas contracts under the 1977 conditions and the oversupply of natural gas. In 1997, Mideast gas was being sold for less than its energy equivalent in oil. Daniel Pearl and Peter Fritsch describe this situation of oversupply in relation to the problems that Qatar's Ras-Gas had in pricing its natural gas:

Since there aren't enough buyers to create a standard price for Mideast gas, contracts are based on the price a buyer would pay for oil producing the same amount of energy. At first, RasGas's agreement with South Korea ensured that the gas wouldn't drop below a price corresponding to $14.44 a barrel of crude. That agreement allowed RasGas to raise

$1.2 billion in a bond offering. RasGas later agreed to scrap the price floor if South Korea, which had reached a deal to buy Oman's gas with no minimum price, would buy twice as much gas. To preserve the RasGas bond rating, Mobil pledged to pay as much as $200 million to bondholders if oil prices fell so low RasGas couldn't meet its bond payments.[64]

Countries also regularly undercut competitors' gas prices to get larger volumes of production. This is done in a manner reminiscent of the many quota violations by OPEC members. As new gas fields are opened, the producers compete for both customers and for financing of their new multibillion-dollar projects. The competition is for the support of multinational utilities, oil producers, banks, investment houses, and for good relations with countries with market and financial clout. All of this adds up to a condition of dependency for the producers of natural gas.

An Argument for Lower Gas Prices

Judith Sack and Robert Chewning[65] argue that without making any heroic assumptions they can predict that gas will be worth as little as 2 cents per kilowatt hour based on gas-fired generation using gas costing $1.50 per million British thermal units. This forecast predicts that natural gas will be in surplus. The forecast also takes into account the structural changes in the natural gas and utilities industries. The reason that the cost of gas is so important is that the price of electricity is generally determined, over time, by the cost of production, and this production, Sack and Chewning predict, will be priced very low because of the low-cost of gas-fired generation.

Beginning with the conventional wisdom that produces a sustainable price of 3.5 cents per kilowatt hour for electricity, Sack and Chewning whittle the price down to a medium term prediction of 2 cents per kilowatt hour. The higher price is based on current technology and current gas prices, but these are expected to change. Over the next several years gas prices are expected to fall because of a surplus of gas and electrical production over demand. Indeed, gas production is expected to produce a bubble of 41 quadrillion cubic feet of gas in known reserves.[66] The longer it takes to produce a sustainable equilibrium price for electricty, the lower the price will fall. Sack and Chewning caution against overinvestment in utilities securities based on the 3.5 cent energy price rather than their much lower 2 cent price.

Sack and Chewning think that the most important factor that will cause falling electricity prices will be the change in the traditional structure of the natural gas market. In addition to these structural changes, they expect an overcapacity in generation equipment to develop because of an attempt to meet the electricity demand of the massive markets in India and China. This excess generating capacity will produce an excess supply of electricity and force down its price. Power marketing will also play a part in producing lower prices, as larger firms cut prices to achieve larger market shares at the expense of price. These same firms will trade energy using old inefficient capacity to meet peak needs (at low prices), while adjusting for customer usage patterns to maximize their output. All of these changes are based on the new convergence between gas and electricity and the new global energy trading practices of multinational utilities that achieve a global reach.

Multinational utilities are expected to trade energy contracts, arbitraging energy and production inputs across national markets to produce the lowest competitive price for electricity. Both gas and electrical capacity will then be monetized and traded along with the other inputs to electrical generation. All of this trading activity will produce new information as energy markets become efficient and markets reach their equilibrium points. The new information will produce electricity at rates as low as available information can make them.

Sack and Chewning also forecast that although new technology will produce plants that run at higher load factors, excess capacity and efficiencies driven by restructuring will last long enough to reduce the sustainable price of electricity, much the same as similar factors did during the 1980s gas bubble.[67]

The (Low) Sustainable Cost of Natural Gas

Because gas-fired electric power generation sets the cost curve for electricity, the most important basis for predicting that the price of electricity will fall to 2 cents per kilowatt hour is the assumption that the price of gas will fall to $1.50 mmbtu.

Traditionally natural gas has been an adjunct to the production of oil, and has been flared because its value was less than the cost of transportation. Because of transportation costs, gas has also traditionally been a highly regulated, regional fuel. The cost of the infrastructure to transport gas—either a pipeline or an LNG facility—has been high relative to the value of the gas to an end user. So natural gas production volumes had to be very large to pay for the capital investments in LNG equipment, pipelines, and the infrastructure necessary to bring it to market.[68]

But new technology is changing the traditional attitudes and structure of the natural gas industry. With new technology, natural gas can be converted economically into both oil and electricity. New energy trading techniques can, in some cases, reduce the transportation cost of gas converted to electricity virtually to zero. International energy arbitrage will stimulate the capital investment necessary to finance gas production with low transportation costs. Under economic conditions of low inflation, gas will also be monetized rather than simply leaving it in the ground to appreciate as prices rise.

A huge global demand for electricity is developing as the growth of global economies—the Asian economies in particular—demand more electricity. Most of this new production will be done through the installation of gas-fired generation because this is the lowest cost technology. The demand for this type of gas-electricity conversion will spur the local infrastructure investments necessary to develop massive amounts of new capacity in natural gas production.

New technologies will bring the new gas production on line economically. New technologies of exploration, imaging, improved directional drilling equipment, and deep water exploration will help increase gas production. Improved catalysts will help the Fischer-Tropsch process convert gas into oil. New gas-fired turbines will improve generational efficiency and the capital costs of the new gas-fired equipment will be recovered in the current, lower-cost electricity as coal-fired equipment becomes obsolete.

Mergers and acquisitions of privatized, formerly state-owned utilities will allow the installation of management teams willing to take risks, and expert enough to make money on the risks taken. These managers will facilitate the rapid adoption of the new technologies in order to increase share price and satisfy mass markets with cheaper power. Stockholder needs will be paramount once private utilities begin to compete in unregulated markets without government financial aid.

All of these changes in the natural gas industry will tend to force down the cost of natural gas and electricity. The expansion of natural gas production in Russia, Iran, Saudi Arabia, and Qatar will tend to provide excess capacity which will push gas prices down further and bring down the cost of electrical generation as well.

Sack and Chewning assume that the electric power generator of the future will get gas at $1.50 mmbtu.[69] Their economic-sized generation firm is a $225 billion 500 megawatt combined cycle gas-fired generation plant with a twenty-year life using MACRS depreciation. The plant will be financed 70 percent with 7.25 percent debt of fifteen years maturity, and the remaining 30 percent will be financed with equity. The plant will pay 35 percent corporate income taxes and have an internal rate of return of 12 percent on after tax cash flows. These assumptions are shown in Table 5.1.

Table 5.1
2 Cent Power—Base Case

Assumptions	
Generating plant	500 MW combined cycle gas
Total Cost of Plant	$225 billion
Capacity factor	95%
Heat rate	6,200 BTUs/kwh
Gas price	$1.50 mmbtu
Book depreciation	30 years, straight line
Tax depreciation	20 years, MACRS
O&M Expense	$0.0040/kwh, escalating annually 2%
Income Tax Rate	35%
Debt financing	$157.50 million (70%)
Equity financing	$67.50 million (30%)
Debt Term	15-yr. final maturity, level amortization
Interest rate	7.25%
Cost of equity	12.0% IRR on after- tax cash flows

Source: Judith B. Sack and Robert L. Chewning, "Global Electricity Strategy: My Two Cents Worth (Or the Sustainable Price of Power)," Morgan Stanley, International Investment Research, February 6, 1997.

An Argument For *Higher* Energy (Gas) Prices

Unexpected Electricity Demand

Some gas industry data imply higher prices rather than the lower prices predicted by Sack and Chewning. The most important factor that might push up electricity rates is an unanticipated demand for electricity. It is the huge global demand for electricity that is now driving gas-to-electricity conversion in the utilities industry. This global demand for electricity is being forced up by the rapid growth of emerging economies financed by their own emerging capital markets, primarily in India, China, the Pacific Basin, and Latin America, while growth in western economies remains robust. The assumption of lower electricity prices assumes that new investment in gas-fired generation will outstrip the world's unprecedented economic growth, and if it doesn't, demand could make electrical rates go up or remain at the 3.5 cent/kilowatt hour level.

The use of old, inefficient generation capacity to satisfy peaks in electricity demand could cut either way, depending on the rate of investment in technologically modern generation equipment. If investment is too fast, as Sack and Chewning suggest, the electrical utilities would tend to produce plenty of lower-cost electricity first by using the modern equipment, forcing electrical rates down. If investment is slower, having to bring on-line a lot of older high-cost generation facilities could keep the cost of electricity above the 2 cent level.

Table 5.2
Gas-to-Oil Conversion Technology

COMPANY	TECHNOLOGY STATUS
Exxon Corp, Dallas	Developing a process that could produce a crude-oil substitute for about $20 per barrel. Plans to build a 50,000 barrel-a-day plant in Qatar.
SYNTROLEUM CORP, Tulsa	Has developed technology suitable for smaller gas fields. The process is economical for producing as little as 5,000 barrels of liquid oil a day. ARCO, Marathon Oil, and Texaco are recent licensees.
SASOL LTD, Johannesburg, South Africa	Now producing diesel fuel from gasified coal The technology was developed during the apartheid-era economic embargo. A joint venture with Norway's Statoil will develop conversion plants for off-shore gas fields.

Source: Company reports, and Gary McWilliams, "Energy: Gas To Oil: A Gusher for the Millennium?" *Business Week*, May 19, 1997, p. 132.

Unexpected Demand for Natural Gas

Unexpectedly high demand for the natural gas to be used in gas-to-oil conversion, by itself, could drive the price of gas up above $1.50 mmbtu. The market for oil is a well-developed, hungry, global consumer market, and the traditional sources of oil are limited and decreasing. This is the reason that Exxon's plans for Fischer-Tropsch conversion plants focus on very high levels of gas-to-oil conversion. Exxon's economic-sized plant would produce up to 100,000 barrels of synthetic crude daily. Such plants would require so much gas, says Exxon CEO Lee R. Raymond, that only "a half dozen or so places" could support them.[70] These plans for gas-to-oil conversion are expected to provide enough oil for almost thirty years.[71] The status of the Fischer-Tropsch conversion technology is shown in Table 5.2.

The Implications of the Cost of Gas

Lower cost gas is good for the consumer, because it would produce lower electricity rates and stimulate productive efficiency. Because low-cost gas stimulates the productive efficiency it stimulates world economic growth as well and is better for everybody in terms of global welfare. Lower natural gas prices would also be good for natural gas producing regions—Russia and the Mideast—because they would be able to sell more of their natural resources in world markets.

Lower cost gas would mean less foreign oil dependence for the United States, and would provoke competition between the regions that supply coal, gas and oil. This means that low-cost gas would pit U.S. coal against Russian and Mideast gas. Although gas is currently touted to become lowest cost source of electrical generation, coal-fired processes may compete well in coal-rich countries like the United States where about half of low-cost generation is coal-fired. Coal technology is improving too. The merger between Louisville Gas & Electric Corporation (LG&E) and KU Energy (KUE) was done to secure for LG&E more coal-fired generation assets to provide electricity at costs that are 50 percent below industry standards.[72]

The probability of low natural gas prices makes investments in electrical utilities risky, because these prices imply lower electrical rates and poor electric utility profits, according to Sack and Chewning. Many of the recent large mergers of so-called "power marketing" U.S. utilities seek to unite gas and electricity in one firm in order to get better access to the lowest cost of energy production.

MERGING TO MARKET NATURAL GAS AND ELECTRICAL POWER

In addition to the economic and technological arguments for gas-to-electricity power mergers, another reason that some companies have merged is to get the partner's experience in marketing gas-to-electricity power products. This was the case in the Enron-Portland General merger.

Another deal that combined many of the key elements which capital markets are scrutinizing in order to determine the value of power companies was the acquisition of Destec Energy by NGC Corporation in February 1997. Destec was 80 percent owned by Dow Chemical Corporation, which had previously sold 20 percent of Destec in order to raise their own share price. The final sale of Destec to NGC, a large national natural gas wholesaler with extensive operations in Texas and California, was done specifically to focus on their core strategies in chemicals and thereby raise the value of their corporation.

Louisville Gas & Electric Corporation and KU Energy Corporation

The $1.43 billion merger of the Louisville Gas & Electric Corporation (LG&E) with KU Energy Corporation (KUE) in May 1997 is a purely domestic example of a merger done to increase market share, operational scale, and efficiency in preparation for national and international energy marketing efforts.[73] KU Energy Corporation specialized in coal-based power generation and did so efficiently enough that it was able to sells electricity at rates nearly 50 percent below the national average.[74] Both of the companies are Kentucky utilities selling gas and electricity.

LG&E is considered to be the more aggressive company, because it has been actively expanding through acquisitions that enlarge their market share and extend their marketing reach from coast to coast. The combined firms would be one of the leading power marketers of gas and electricity in the United States, with revenues of $4.3 billion and 1.1 million customers. Despite the fact that LG&E is Kentucky's largest utility, it has only a modest home market. "This [merger] creates one of the largest low-cost energy holding companies in the country,"said Roger W. Hale, the chairman and chief executive of LG&E.[75]

The merger would have to get the approval of five state and federal agencies, including the FERC, but this approval was likely because of the trend to allow mergers to help utilities compete more effectively in the rapidly deregulating U.S. power generation industry.

In the short run, LG&E planned to reduce their dividends by 2 cents a share but said that the merger would save them $760 million over a period of 10 years. The merged company would reduce their work force by about 400 through attrition, leaving them with about 5,000 employees. LG&E planned to reduce electricity rates about 2 cents per kilowatt hour and promised not to raise rates unless they were faced with extraordinary circumstances.[76]

LG&E paid for controlling interest in KUE by swapping 1.67 shares of their stock for each share of KUE stock to give KUE a 30 percent premium on their shares.[77] KUE shares closed up $4.125, or 14 percent, with KUE shareholders receiving LG&E shares worth $37.78 for each KU share.[78] LG&E was considered to have paid more than average for KUE. "They have to be bigger," said Edward J. Tirello, an analyst at Natwest Securities, one of several who predicted such a combination. "And the one you know best is your neighbor."[79]

Hale, LG&E's CEO, recently pulled out of the Edison Electric Institute (an industry lobby group) because of their reluctance to support gas and electricity deregulation. LG&E is a low-cost producer, like many utilities who have sought rapid deregulation,[80] and this probably explains Hale's withdrawal. Companies seeking to delay deregulation are thought to have higher costs, often due to billions of dollars of investment capital stranded in high-cost nuclear power plants.

Enron and Portland General Corporation

In July 1996, Enron Corporation of Houston announced a $2.1 billion buyout of Portland General Corporation of Oregon.[81] Enron is the nation's largest wholesaler of natural gas, and it has enthusiastically embraced the concept of deregulation. Enron's strategy has been to compete actively in deregulated markets and to do this effectively, it has been developing unique markets and services in natural gas. It plans to do the same thing in electricity. One way it has distinguished itself in the marketing of natural gas is through ingenious pricing formulas. For example, Enron pegged the price of natural gas for Florida citrus producers to the price of juice futures. If juice prices fell, so did a grower's gas fuel costs.

Enron also intends to become a nationwide one-stop seller of both natural gas and electricity. Their $3.2 billion buyout of Portland General, unified the large Oregon electrical power company with their own, non-contiguous, Texas-based business in natural gas. Under its new rules, FERC expedited their approval of this merger,[82] and the merger set the industry pattern for U.S. gas-electric combinations. Enron is currently testing retail sales of gas and electricity in New Hampshire and New York, two of the first states to open their residential power markets to newcomers.

Diverse Technology: Wind Turbines

Enron is a world power marketer of not only natural gas and gas-fired electricity but also of electricity generated by wind turbines. While Enron's core product is electricity, it has diversified the technologies with which it produces the product—diversification through technological innovation. In March 1997, Enron agreed to build 150 propellered wind turbines in Iowa for a cost of $100 million. The turbines will generate 112.5 megawatts of electricity which would supply about 50,000 homes.[83] This is the world's largest wind-generated electricity project and the source of power is cheap, environmentally clean, and renewable.

Enron will build the wind turbines for Iowa's MidAmerican Energy to satisfy federal regulations that require it to seek alternative power sources.[84] Standing on masts about fifteen stories tall, the turbines' 25-ton fiberglass and epoxy rotors will have about the same wing span as a Boeing 747. Over the last ten years, the cost of wind-generated electricity has fallen to about 4 to 4.5 cents a kilowatt hour, but this is still not as cheap as 3.5 cent gas-fired electric generation.[85] Aside from federal regulations, wind-generated power can be desirable because it is relatively cheap, renewable, and non-polluting, and it is cheaper than other alternatives such as solar power. "As customer choice arrives, that will be good for renewables," said Randall Swisher, executive director of the American Wind Energy Association.[86]

Marketing Strategies

Enron's U.S. strategies are colored by tough negotiating attitudes honed by negotiating foreign construction projects. According to Enron's Rebecca P. Mark, you have to be pushy and aggressive if you are going to be successful in India's utilities industry. Mark was responsible for rewinning Enron's canceled $2.5 billion power project in India. To do this Enron had to play Indian politics for five years to get the approval of three successive Indian governments, reverse the latest cancellation, and win twenty-four lawsuits.[87]

Some of the perceptual barriers that Enron overcame with five years of virtually continuous negotiations were the following :

* The large project was perceived as too grandiose.
* Enron took a local Indian partner when Indian joint ventures were thought to "go nowhere."
* India's huge bureaucracy thought Enron was too pushy and aggressive.
* The project was high priced, but Enron had to price the risk into the project.[88]

Now Enron has finally begun construction of a large, modern liquefied natural gas power plant. Mark is currently leading negotiations for an additional $10 billion power investment.

In June 1997, Enron announced its forthcoming entry into the Italian market through a $2.94 billion joint venture with Enel, an Italian state electricity group. The deal would be a 50-50 joint venture with approximately 80 percent debt financing. The remainder of the financing, after the $2.4 billion of debt, would consist of approximately $600 million in equity stakes to be shared equally by Enron and Enel. Enel would transfer generating plants with up to 5,000 megawatts generating capacity into the group, and, as a part of the agreement, Enron would upgrade these single-cycle oil and gas-fired generating plants to a combined cycle gas-fired plants.[89] In return for its stake, Enron would get a 50 percent interest in the venture and access to the European Common Market under the market's electricity liberalization program.[90]

In order to satisfy the Common Market program, Enel would have to divest themselves of 30 percent of their domestic electricity monopoly by 1999. By contributing the plants to the Enron-Enel joint venture, Enel would get to keep the benefits of these plants and simultaneously satisfy the liberalization program. Enron and Enel, the Italian oil and gas group, have just concluded a similar agreement.[91]

Enron is already a leading energy power marketer in Turkey and Sardinia, and hopes to enter the $200 billion European retail market for electricity with its Enel deal.[92] Enel needs Enron's risk management and marketing expertise to help it deliver power and compete effectively in European markets which will be deregulated by 1999. This would be the first U.S. equity investment in an Italian utility and closely follows the Southern Company's deal to buy 25 percent of Berlin's electric utility, Berliner Kraft & Licht AG, for about $830 million.[93]

Table 5.3
Two Large Gas-to-Electric Mergers Set the Pattern

Announced	Acquiring Firm	Target Firm	Value
Jul 1996	Enron Corporation	Portland General	$2.1B
Nov 1996	Duke Power Company	PanEnergy	$7.7B

Source: Benjamin A. Holden, "Enron Corp. Has Accord to Buy Portland General," *Wall Street Journal*, July 22, 1996, p. A3; Sullivan Allanna, "Enron Deal Signals Trend in Utilities," *Wall
Street Journal*, July 23, 1996, p. A3; Allen R. Myerson, "Enron Will Buy Oregon Utility
in Deal Valued at $2.1 Billion," *New York Times*, July 23, 1996, p. D1; and Steven Lipin
and Peter Fritsch, "Duke Power Plans to Acquire PanEnergy In Stock Transaction of
About $7.7 Billion," *Wall Street Journal*, November 25, 1996, pp. A1, A3.

Duke Power Company and PanEnergy Corporation

In November 1996, Duke Power Company announced plans to acquire the
PanEnergy Corporation for $7.7 billion in order to create a huge network of
utilities to market natural gas and electricity. The deal signaled another large regional power producer's strategy to play to their customer's increased ability to
bypass their local utility and buy directly from low-cost energy marketers.[94] The
deal was very similar to the combination of Enron and Portland General, but unlike most other recent mergers which had attempted merely to extend customer
bases and reduce costs.

The huge Duke-PanEnergy merger, following that of Enron-Portland General, began to establish the gas industry as the merger model for the more slowly
adapting electrical power industry, and is by far the most significant recent transaction in the U.S. utility merger market. This merger, like that of Enron-
Portland, attempted to create one-stop energy shopping for businesses and
consumers in the southeast. The merger was also similar to the merger of
Houston Industries, Inc., which announced its intention to acquire NorAm
Energy Corporation for $2.4 billion in November of 1995.[95] The two largest
gas-electric mergers are shown in Table 5.3.

NGC Corporation And Destec Energy Corporation

In February 1997, NGC Corporation announced plans to acquire Destec
Energy Corporation, a utility specializing in electrical power generation, for
$1.27 billion.[96] NGC is a rapidly growing Houston-based company that sells
natural gas. Destec was 80 percent owned by Dow Chemical Company. NGC
was primarily interested in acquiring Destec's U.S. electricity generation plants
to increase its foothold in U.S. electricity markets and plans to sell Destec's international operations to AES Corporation of Arlington, Virginia, another power
producer, for $365 million.[97]

In 1984, NGC began building up its natural gas brokerage business from a
Houston-based subsidiary called the Natural Gas Clearinghouse. They expanded
this business as the industry became deregulated and quickly grew into one of

the largest gas brokers in the United States. NGC has been tripling its wholesale electricity business every two to three months by supplying large industrial customers. NGC pioneered selling low-cost gas to large out-of-state customers, with regulatory approval, even before the deregulation of U.S. energy markets, and continues to prefer this kind of customer.[98]

Destec had been on the market for a while, its stock had been flat, although NGC paid an estimated 11 percent premium for the company.[99] Dow had first considered spinning off Destec in October 1996 as a means of raising their stock price, which it had already begun by selling a 20 percent stake to the public in 1991.[100] Destec had recently lost large contracts with Texas Utilities (a unit of Houston Industries Inc). After these losses it had hired Morgan Stanley to explore its strategic alternatives.[101] Dow's spinoff of Destec recognized the increasing degree of specialization and expertise necessary to manage a competitive gas-to-electric utility.

On the announcement of the deal with NGC Destec's shares gained 4.8 percent while NGC's shares slipped six-tenths of a percent. The part of the deal that the stock market like best was AES's acquisition of Destec's overseas assets in markets that were new to them, like their investments in the Netherlands and the Dominican Republic.[102] NGC also planned to sell Destec's coal and oil reserves, a coal gasification plant, and a Florida power plant[103] AES's stock went up 7.6 percent to $66.75 on the announcement of the deal. Destec's international power assets are shown in Table 5.4.

Chuck Watson, NGC's chairman and chief executive, said that with Destec's assets, NGC "becomes a major player in the power-generation business."[104] Like Enron, NGC is an established natural gas marketer, which hopes to use its natural gas marketing infrastructure to sell Destec's electricity in a deregulated market—a copycat oligopolist, or a fledgling power-marketer.[105]

Table 5.4
Destec's International Electrical Power Assets

Announced	Acquiring Firm	Target Firm	Value
Nov 1995	Destec Energy Inc. electricity and gas	55 % of Dominican Republic Power Plant project from Turbine Energy Inc.	$100M
Aug 1996	National Power PLC (U.K.), Commonwealth Bank Group (Australia), and PacifiCorp (U.S.), Destec Energy Inc. (U.S.)	Hazelwood Power Station (Australian)	$1.86B

Sources: "Destec Energy Inc.: Power Plant Project Stake Acquired for $100 Million" *Wall Street Journal*, November 24, 1995, p. D4; and Benjamin A. Holden "PacifiCorp and Destec Join Group to Pay $1.86 Billion for a Utility in Australia," *Wall Street Journal*, August 5, 1996, p. A4.

Destec is a good base from which to expand NGC's power-marketing of electricity because Destec is a source of low-cost electricity. Destec produces 1,800 megawatts of this power in the twenty gas-fired plants it operates in the United States.[106] Eleven of Destec's plants are located in California where deregulation is proceeding more rapidly than elsewhere in the U.S.. The other plants are located in Texas, Florida, Georgia, Virginia, Michigan, and Indiana.[107]

"What is driving the trend toward consolidation is that states like California are preparing to [rapidly] open their electricity markets to competition. Companies like NGC, Enron and PanEnergy are setting up nationwide marketing arms in attempts to win electricity and natural gas customers from local utilities."[108] One reason for NGC's buying Destec in February, was that they expected quick regulatory approval from FERC—in months rather than years—because of FERC's announcement of a set of new fast approval guidelines.[109] Unlike Portland General, that was recently acquired by Enron, Destec was an independent power producer which was not a regulated.[110]

PacifiCorp and Energy Group PLC

PacifiCorp was already an electricity and coal power marketer before it announced its plans to acquire the U.K.'s Energy Group PLC for $5.8 billion in cash in June 1997.[111] PacifiCorp offered a premium for the utility not only because they wanted access to Energy Group's coal assets, but because they also wanted an international position from which to arbitrage coal and electricity assets internationally. PacifiCorp already sold low-cost electricity well beyond the seven Western states where it had 1.4 million customers.

PacifiCorp's U.S. strategy was to buy cheap raw coal in the Midwest, convert it into cheap electricity, and sell the electricity in high-cost power areas such as the east coast. Their bid for Energy Group signaled their intention to extend the range of this strategy by arbitraging coal and electricity internationally between England and the United States. PacifiCorp would transact this arbitrage by trading coal and energy contracts developed with their customers in the different markets—and different countries—that they serve.

Although they are both power marketing utilities, Enron and PacifiCorp operate in different markets—Enron in the gas-to-electricity market, and PacifiCorp in the coal-to-electricity market. Although gas-fired electricity generation is thought to be the lowest cost generation method of the future, at the present time, rich U.S. deposits of low-cost coal make companies like PacifiCorp and LG&E lower cost generators than the gas-fired power producers. PacifiCorp, however, has the ability to arbitrage coal, gas, or electricity. PacifiCorp can sell a customer whichever product they want as long as they can make a profit on it. This gives the customer, whether it is a business, a broker, or a utility, the ability to choose the lowest cost product. By deregulating its utilities, the United States is giving energy customers the power to choose suppliers nationwide, and many of these suppliers, like PacifiCorp, will be operating internationally.

International Cross Investment

Countries like Great Britain, which have already privatized and deregulated their utilities, offer American power marketers important advantages when they allow U.S. foreign investors to buy into their formerly state-owned utilities. One indication of how valuable it is for a U.S. utility to have a foothold in a foreign market is the fact that U.S. utilities have bid $21.436 billion for British and Australian utilities in the period from July 1995 to June 1997.[112] Free-marketeering Britain is eager to have the best possible management of their utilities to keep their production costs low, their share values high, and keep electricity customers happy with low rates. This did not happen when Britain's public utilities were operated in regulated markets. But now British policy has implemented both deregulation and privatization. These policies are thought to be the cure for inefficiency in power production and since Britain has taken this prescription, energy rates have fallen.

The process of direct foreign investment is not a one way street. Many large U.K. generators have invested abroad heavily for exactly the same reasons that U.S. firms have invested in Britain. At least two large British utilities have had their domestic acquisitions blocked because of the government's fear that they would monopolize and dominate Britain's fledgling power markets. Having been refused acquisitions in the U.K., British PowerGen invested £700 million ($1.12 billion) in Asia, Australia, Europe, and the United States. When the same thing happened to National Power, they invested £1 billion ($1.6 billion) abroad.[113]

Cross investment is turning multinational utilities into global utilities, and the international energy arbitrage of these global utilities is simultaneously developing an integrated global energy market. If the traders both operate in the two countries involved, energy arbitrage can often dramatically reduce both the costs and risks of energy production by allowing contracts to be traded rather than the commodities themselves. The ideas that make this possible for multinational companies are the concepts of modern market economics, privatization and deregulation programs, pushed forward by the new technologies of energy convergence. The natural tendency of this process will be to apply the law of one price to global energy supplies. All different energy products will tend to be sold in units whose prices are equivalents to the lowest competitive price of the most efficient power source. The technologies of energy convergence will implement this price by allowing the conversion of coal, gas, and oil into electricity and the conversion of gas into oil, or vice versa.

Western Resources, ADT and Tyco

The mergers and acquisitions activity between Western Resources, ADT, and Tyco illustrates how the hypothetical synergies of some mergers strain one's credibility.

Western Resources Inc. is a large, midwestern energy utility which expanded recently by completing its acquisition of Kansas City Power and Light Company (KCP&L) in a bitterly fought $2 billion takeover. In 1996 Western made a $4.3 billion hostile bid for 28 percent of ADT, the world's largest producer of home security systems. Following that, Western bought Westinghouse Security Systems for $368 million and then proceeded with a hostile stock and cash bid for the 73 percent of ADT it did not own. Western apparently wanted to become a one-stop power marketer by combining its sales of residential energy products with sales of other home security products. Unlike the mature, increasingly competitive energy business, the security business offered double-digit growth rates and a market that remained largely unexploited.

Unfortunately for Western, ADT did not see any significant synergies between home security products and electrical power. So, in March 1997, ADT announced a $5.4 billion friendly bid by Tyco, which topped Western's hostile $4.3 billion bid, thwarting the utility. "With Tyco, you get fire and security, which is much more compatible than a utility," said Scott Lowry, an analyst with Portsmouth Financial Services in San Francisco.[114] "This deal is the difference between a bulls eye and a black eye," said Michael A. Ashcroft, chairman of ADT. "It's the merger of fire protection and security services. There aren't really any synergies with an electric utility based in three states in the Midwest and the national structure we've got."[115]

The tenacity of Western Resources illustrates how entrenched attitudes about the compatibility of market-mate energy products can become. Carl Koupal, a spokesman for Western, said the company was "taking a close look at the true value of the Tyco offer" before issuing a response.[116] At the very least, Lowry commented, the Tyco offer represented "a more realistic market value" for ADT, than what he called a "low ball" bid from Western. "But I think Western has a little room to continue to play this game. I wouldn't be surprised if we see a response from Western. I think they want it," Lowry said.[117]

The markets liked the Tyco bid and ADT shares soared nearly 18 percent to $25.625. Tyco International lost $2, to $58.25. Based on that price, the merger was worth $28.03 a share, or $5.4 billion. Western shares closed at $31, up 87.5 cents. Now Western has a choice of either raising its bid for ADT or taking a $600 million capital gain on its year-old investment in the ADT. The ADT-Tyco combination would create a leader in both fire protection and commercial and home security with more than $8.5 billion in annual revenue.[118]

Information and Computers

Underlining the importance that they attach to energy transactions in privatized energy markets, in December 1966 Enron named Jeffrey K. Skilling as its number two executive. Skilling was known for building up a huge business for Enron in a subsidiary that specialized in trading natural gas and electricity. The move showed how quickly the trading and marketing of gas and electricity had grown in importance because of utility deregulation and how fast U.S. utilities were scrambling to keep up.[119]

Kenneth L. Lay, CEO of Enron feels that a good computer system is crucial to success in power marketing and he thinks that Enron owns such a system. It cost Enron approximately $250 million to develop its program, but Lay estimates that good information processing in today's competitive markets for electricity and gas could squeeze as much as 30 to 40 percent out of their current costs.[120] Enron's programs for marketing and trading handle complex electricity trades and energy arbitrage. Lay predicts that most of the savings will show up in rate cuts for consumers.[121]

NEW WHOLESALE MARKETS FOR POWER

Mergers are not the only way that utilities have restructured themselves to manage energy more efficiently. Energy alliances such as that between Oglethorpe Electric and Louisville Gas & Electric Company (LG&E) have hedged the risk of energy marketing by contracting out part of a utility's production. Enron settlement of their long-term contract loss on North Sea gas illustrates the kind of unhedged risk that gives energy companies nightmares. The merger of PacifiCorp and Energy Group PLC illustrates how global utilities can spread these risks internationally by building up multinational trading facilities. This section ends with a description of "freebooters," corporations that are not utilities, but simply companies that craft and trade energy contracts.

Energy Alliances

In January 1997, Enron teamed up with eleven California cities in an energy alliance to provide gas, operational management, risk management, and other financial products such as securitized energy.[122] The products would be supplied to 700,000 customers through its agreement with the Northern California Power Agency (NCPA). Twenty-five percent the electricity in the United States is provided by smaller municipal agencies like NCPA. These agencies are often not-for-profit public corporations that face increased competition from larger utilities as deregulation proceeds. If the agencies lose customers, they will have less revenue to pay for other municipal services such as garbage pickup and road repair.[123]

The alliance is a defensive move by NCPA to protect its business from Pacific Gas & Electric Co., a giant California electric utility which currently provides natural gas to NCPA customers and is expected to compete with NCPA for electricity customs as deregulation evolves.[124] The Enron alliance gives NCPA another potential gas provider to compete with PG&E as the state of California moves quickly to open all its energy markets to competition from both within and without the state. Enron intends to begin spending $200 million a year in 1997 in a nationwide advertising campaign to sell its corporate image along with the idea of cheaper energy in a deregulated world.[125]

Enron's marketing strategy is to provide local utilities and agencies with power rather than competing with them for their customers.[126] This strategy helped Sprint and MCI compete with AT&T after its breakup. The Northern California agency is a good place to test this strategy, because the NCPA is a government monopoly looking for outside private assistance in serving its customers. It is also looking for new energy management ideas and economies of scale of management. The agency has electric rates that are 50 percent higher than the national average, whereas Enron is a low rate national power marketer of both gas and electricity.[127] The Enron-NCPA deal may serve as a model for other public-private partnerships in a deregulated electricity market.

Electric Utilities Contract Out for Electricity

Georgia-based Oglethorpe Power's deal with LG&E demonstrates another strategy for limiting the risk of competition. Oglethorpe is a large regional utility that is finding itself becoming a small national utility as U.S. mergers consolidate the industry.

Seeking big cost cuts and reduced risk, in November 1996 Oglethorpe turned to LG&E, an out-of-state power-marketing company, to supply half of its electricity needs. This deal was the largest long-term power contract yet made, a contract for $4.5 to $5 billion of electricity, producing tens of millions of dollars of profit and extending for over a decade.[128] The contract was an example of the trend among power marketers to shop for good energy prices even when they were producing that commodity themselves. The deal also demonstrated the new ability of electricity producers to negotiate and take on long-term contracts. LG&E, which is expected to supply Oglethorpe with over half its power for fifteen years, is also negotiating long-term power contracts with Enron and Duke/Dreyfuss.[129]

Oglethorpe supplies power to thirty-nine electric cooperatives serving 2.6 million households, yet it feels that it could not match the marketing skill and large volumes of LG&E, and felt that it had less experience purchasing power than LG&E.[130] LG&E, on the other hand, is a coal-fired generator of low-cost electricity, and one of the most aggressive utilities to market power outside its home state. LG&E routinely buys and sells massive amounts of power daily, from many sources.

Oglethorpe will buy electricity from outside sources if that power is cheaper than their own production. It will also sell power from its own plants if their production is lower cost than LG&E's. Oglethorpe's contract has locked in a certain price that LG&E must meet. Oglethorpe can buy elsewhere if LG&E does not meet that price. If LG&E can buy power below Oglethorpe's price, it can sell to Oglethorpe and keep the difference. This is much like LG&E's *buying a put* on electricity prices. In effect, LG&E bets that their prices will fall or remain the same so that they will profit from the difference between the contract price at which they will supply Oglethorpe and the lower market price. Oglethorpe, on the other hand, in effect, *sells a put*. Selling a put has the same effect as buying a call: it is Oglethorpe's bet that their energy prices will rise above LG&E's supply price. This type of contracting plainly focuses on *risk management*. It becomes important that power producers carefully follow market prices, forecast supply and demand, and hedge some of the risks implicit in energy price movements. What's new in all of this is that Oglethorpe has decided to let outsiders—i.e., the market—handle a large part of their power production.

Sometimes Even the Big
Players Get the Price Wrong

Like LG&E, Enron markets gas and electricity outside its home state (Texas). But all of Enron's energy deals are not profitable. In order to make high profits Enron must manage the high risk of global energy markets. A lot of Enron's risk is bound up in their long-term contracts to receive or deliver power at fixed rates. These contracts are the equivalent of speculative gambles on the directions that energy prices will take in the future. The values of these contracts fluctuate with the market prices of the energy products that they are based on and behave much like risky commodities futures contracts.

In June 1997, Enron negotiated a $440 million settlement with Phillips Petroleum Company over its long-term contracts for delivery of North Sea gas. Enron had made these contracts to assure themselves a supply of fuel for two new gas-fired electric power plants it was building.[131] They agreed to pay Phillips and its partners more than $3 dollars per thousand cubic feet of gas, hoping to sell the gas to their new plants in the U.K.[132] When Enron entered the contract in 1993, it intended to buy about 800 billion cubic feet of gas. But the construction of the power plants was delayed and then natural gas prices fell, leaving Enron with a $675 million loss on the contracts, which equaled about $1.80 (4.5%) a share.[133] Because of the circumstances, Enron refused to accept delivery of the gas in September 1996—they had no power plants in which to burn it—and renegotiated the prices of the contracts in June 1997.[134] Enron said that the renegotiated price for the gas would be "significantly less" than the originally contracted price.[135]

This is an example of the high risk that power-marketing global utilities face when they deal in the products of gas-to-electricity conversions. To get out of the contract, Enron paid Phillips $440 million and wrote off the remaining $235 million as a quarterly loss on its income statement.

Freebooters

The same type of contracts and pricing strategies used by wholesale energy suppliers like LG&E will work for freebooters, marketers who own neither wires nor power plants.[136] These entrepreneurs also use computer systems to craft complex energy contracts and create customized pricing plans, and are expected to package electricity with other products like telephones. These power marketers will use their competitor's wires to deliver electricity at a lower price, much as the Bell System was forced to let MCI use their local phone lines. Freebooting can be lucrative in an energy market which has an estimated $150 billion over-supply (McKinsey & Co.) of electrical generating plants. The freebooters will make their profit by chipping away at customer bases where regulation has propped up high prices.

NOTES

1. Benjamin A. Holden, "PacifiCorp Pursues Energy-Swap Plans," *Wall Street Journal*, June 16, 1997, pp. A1 , B4.

2. *Ibid.*

3. Benjamin A. Holden, "Enron Corp. Has Accord to Buy Portland General," *Wall Street Journal*, July 22, 1996, p. A3; Sullivan Allanna, "Enron Deal Signals Trend in Utilities," *Wall Street Journal*, July 23, 1996, p. A3; and Allen R. Myerson, "Enron Will Buy Oregon Utility in Deal Valued at $2.1 Billion," *New York Times*, July 23, 1996, p. D1.

4. Steven Lipin and Peter Fritsch, "Duke Power Plans to Acquire PanEnergy in Stock Transaction of About $7.7 Billion," *Wall Street Journal*, November 25, 1996, pp. A1, A3.

5. Agis Salpukas, "CalEnergy Offers to Buy British Utility," *New York Times*, October, 19, 1996, pp. D1, D7; "British Utility Lifts Payout in Face of Bid," *New York Times*, December 11, 1996, p. D6; and "CalEnergy Bid For Utility is Victorious. British Distributor is Latest Acquisition," *New York Times*, December 25, 1996, pp. D1, D3.

6. James P. Miller and Steven Lipin, "CalEnergy Launches Another Hostile Bid," *Wall Street Journal*, July 16, 1997, pp. A1, A3, A4; and Agis Salpukas, "$1.9 Billion Hostile Bid for Utility," *New York Times*, July 16, 1997, pp. D1, D18.

7. Gary McWilliams, "Energy: Gas to Oil: A Gusher For the Millennium?." *Business Week*, May 19, 1997, p. 132.

8. *Ibid.*

9. *Ibid.*

10. *Ibid.*

11. *Ibid.*

12. Daniel Pearl and Peter Fritsch, "Deep Pockets: Natural Gas Generates Enthusiasm and Worry in Oil-Soaked Mideast," *Wall Street Journal*, August 11, 1997, pp. A1, A8.

13. *Ibid.*

14. *Ibid.*

15. *Ibid.*

16. *Ibid.*

17. *Ibid.*

18. *Ibid.*

19. *Ibid.*

20. Peter Fritsch and Maureen Kline, "Enron, Italian Utility ENEL Expected To Announce Power-Marketing Venture," *Wall Street Journal*, June 3, 1997, pp. A1, A3.

21. *Ibid.*

22. Pearl and Fritsch, "Deep Pockets," pp. A1, A8.

23. Bloomberg News, "ARCO to Develop Gas Field in Indonesia," *New York Times*, September 3, 1997, pp. D1, D7.

24. McWilliams, "Energy: Gas To Oil," p. 132.

25. *Ibid.*

26. *Ibid.*

27. *Ibid.*

28. *Ibid.*

29. *Ibid.*

30. *Ibid.*

31. *Ibid.*

32. *Ibid.*

33. *Ibid.*

34. *Ibid.*

35. *Ibid.*

36. *Ibid.*

37. *Ibid.*

38. *Ibid.*

39. Reuters, "Russian Utility to Try to Raise at Least $2 Billion," *New York Times*, May 17, 1997, pp. D1, D4.

40. Andy Reinhart, "The Business Week Global 1000: The Russians Are Here, The Russians Are Here," *Business Week*, July 7, 1997, pp. 94-96.

41. Reuters, "Russian Utility to Try," pp. D1, D4.

42. Reinhart, "The Business Week Global 1000," pp. 94-96.

43. *Ibid.*

44. *Ibid.*

45. *Ibid.*

46. *Ibid.*

47. Judith B. Sack and Robert L. Chewning, "Global Electricity Strategy: My Two Cents Worth (Or the Sustainable Price of Power)," research paper, Morgan Stanley, International Investment Research, February 6, 1997.

48. *Ibid.*

49. *Ibid.*

50. Reinhart, "The Business Week Global 1000," pp. 94-96.

51. Pearl and Fritsch, "Deep Pockets," pp. A1, A8.

52. Sack and Chewning, "Global Electricity Strategy."

53. McWilliams, "Energy: Gas to Oil," p. 132.

54. Holden, "PacifiCorp Pursues Energy-Swap Plans," pp. A1, B4.

55. *Ibid.*

56. McWilliams, "Energy: Gas To Oil," p. 132.

57. Pearl and Fritsch, "Deep Pockets," pp. A1, A8.

58. Holden, "PacifiCorp Pursues Energy-Swap Plans," pp. A1, B4.

59. Sack and Chewning, "Global Electricity Strategy."

60. *Ibid.*

61. McWilliams, "Energy: Gas to Oil," p. 132.

62. *Ibid.*

63. Pearl and Fritsch, "Deep Pockets," pp. A1, A8.

64. *Ibid.*

65. Sack and Chewning, "Global Electricity Strategy", McWilliams, "Energy: Gas to Oil," p. 132, Pearl and Fritsch, "Deep Pockets," pp. A1, A8; Gundi Royle, "Gas in Europe—The Door to Competition is Pried Open," *Monthly Perspectives*, May 1996, and Richard H. K. Vietor, *Contrived Competition*, New York: Morgan Stanley.

66. McWilliams, "Energy: Gas to Oil," p. 132.

67. Sack and Chewning, "Global Electricity Strategy."

68. Pearl and Fritsch, "Deep Pockets," pp. A1, A8.

69. Sack and Chewning, "Global Electricity Strategy."

70. McWilliams, "Energy: Gas to Oil," p. 132.

71. *Ibid.*

72. Benjamin A. Holden, "LG&E to Buy KU for $1.43 Billion in Stock," *Wall Street Journal*, May 22, 1997, p. A4.

73. *Ibid.*

74. *Ibid.*

75. Allen R. Myerson, "LG&E Energy Agrees to Buy Rival Utility in Kentucky," *New York Times*, May 22, 1977, pp. D1, D21.

76. *Ibid.*

77. *Ibid.*

78. Holden, "LG&E to Buy KU," p. A4.

79. Myerson, "LG&E Energy Agrees to Buy," pp. D1, D21.

80. *Ibid.*

81. Gary McWilliams, "Enron's Pipeline into the Future," *Business Week*, December 2, 1996, p. 82.

82. Agis Salpukis, "Enron Utility Merger Approved in Move to Expand Nationwide," *New York Times*, February 27, 1997, pp. D1, D4.

83. Allen R. Myerson, "Enron Wins Pact to Supply Power from Wind Turbines," *New York Times*, May 27, 1997, p. A5.

84. *Ibid.*

85. *Ibid.*

86. *Ibid.*

87. Manjeet Kripalani, "India: 'You Have To Be Pushy and Aggressive", *Business Week*, September 24, 1997, p. 56.

88. *Ibid.*

89. Paul Betts (Milan), "Enel, Enron Set to Form Joint Venture," *Financial Times*, June 4, 1997, p. 15.

90. *Ibid.*

91. Fritsch and Kline, "Enron, Italian Utility ENEL Expected," pp. A1, A3.

92. *Ibid.*

93. *Ibid.*

94. Agis Salpukas, "A 7.7 Billion Union of Gas, Electricity, Duke Power Gaining PanEnergy's Sales Skill," *New York Times*, November 26, 1996, pp. D1, D6.

95. Lipin and Fritsch, "Duke Power Plans to Acquire PanEnergy," pp. A1, A3.

96. Carlos Tejada, "NGC to Acquire Destec for $127 Billion. Natural gas Concern Aims to Stake Out Position in Electricity Market, " *Wall Street Journal*, February 19, 1997, pp. A1, A2.

97. Agis Salpukas, "Growing Natural gas Seller to Expand Electric Business. NGD Buying Destec, a Plant Operator," *New York Times*, February 19, 1997, pp. D1, D2.

98. *Ibid.*

99. Tejada, "NGC to Acquire Destec," pp. A1, A2.

100. Salpukas, "Growing Natural gas Seller," pp. D1, D2.

101. Tejada, "NGC to Acquire Destec," pp. A1, A2.

102. *Ibid.*

103. *Ibid.*

104. *Ibid.*

105. *Ibid.*

106. Salpukas, "Growing Natural gas Seller," pp. D1, D2.

107. Tejada, "NGC to Acquire Destec," pp. A1, A2.

108. Salpukas, "Growing Natural gas Seller," pp. D1, D2.

109. *Ibid.*

110. Carlos Tejada, "NGC to Acquire Destec," pp. A1, A2.

111. Agis Salpukas, "PacifiCorp is Said to Reach Deal to Buy British Utility," *New York Times*, June 12, 1997, pp. D1, D7.

112. From Table 5.1 in Chapter 5, British and Australian Utilities Mergers.

113. Simon Holberton, "Generators Plug in Abroad," *Financial Times*, May 26, 1997, p. 17.

114. Charles V. Bagli, "ADT and Tyco Plan to Merge in $5.4 Billion Stock Swap," *New York Times*, March 18, 1997, pp. D1, D21.

115. *Ibid.*

116. *Ibid.*

117. *Ibid.*

118. *Ibid.*

119. *Ibid.*

120. Agis Salpukas, "Enron Names Trading Chief as President of Company, Appointment Reflects the Changing Market," *New York Times*, December 11, 1996, pp. D1, D6.

121. *Ibid.*

122. Peter Fritsch, "Enron to Unveil Energy Alliance Involving 11 Cities in California," *Wall Street Journal*, January 15, 1997, pp. A1, B4.

123. *Ibid.*

124. *Ibid.*

125. *Ibid.*

126. *Ibid.*

127. *Ibid.*

128. Agis Salpukas, "Utility Seeks Partner to Aid Power Needs," *New York Times*, November 20, 1996, pp. D1, D2.

129. *Ibid.*

130. *Ibid.*

131. Bloomberg News, "Enron to Pay $440 Million to Settle Gas Dispute," *New York Times*, June 3, 1997, pp. D1, D6.

132. Fritsch and Kline, "Enron, Italian Utility ENEL Expected," pp. A1, A3.

133. *Ibid.*

134. Bloomberg News, "Enron to Pay $440 Million," pp. D1, D6.

135. Fritsch and Kline, "Enron, Italian Utility ENEL Expected," pp. A1, A3.

136. Peter Coy and Gary McWilliams, "Electricity: The Power Shift Ahead," *Business Week*, December 2, 1996, pp. 78-82.

Chapter 6

Nuclear Power

BACKGROUND

During the 1950s and 1960s the United States tried to beat swords into plow-shares and built its cohort of nuclear electricity generating plants to create a peaceful use for atomic energy. So many of these plants were built that, in retrospect, it is clear that the United States seriously overinvested in this expensive source of power production.[1] It is also generally acknowledged now that nuclear power generating technology has not lived up to its original promise.

This somber conclusion had not been reached in the 1960s, which was a period of construction and optimism. During those years, the nuclear power industry was dominated by issues involving the source and cost of uranium fuel. Uranium from the United States was more expensive than uranium from the Soviet Union, but domestic supplies were encouraged to protect the fledgling U.S. industry. This was done primarily because uranium was still needed for nuclear warheads and a domestic source was preferred during a time when the United States was building up its nuclear arsenal and the Cold War was still running strong.

During the 1970s, advocates of atomic energy hoped that development of the breeder reactor would provide more abundant, lower cost nuclear fuel. The nuclear reactions in a breeder reactor were expected to produce a cascade of by-products most of which could, themselves, be used as nuclear fuel. But the breeder reactor never worked, and, even worse, there is currently no satisfactory solution to the storage of nuclear wastes. These wastes include plutonium, which has a very long halflife and is one of the most toxic cancer-producing substances ever known.

In 1997, far from having an uranium shortage, the U.S. government has more nuclear fuel than it needs. Nuclear warheads are being decommissioned and the president has even encouraged congress to appropriate money to buy the nuclear parts of decommissioned Russian warheads. The U.S. Enrichment Corporation (USEC), a federal corporation located in Bethesda, Maryland, converts these warheads into nuclear fuel that can be used by U.S. nuclear generation plants under an $8 billion deal with Russia.[2] The material in these discarded Russian warheads is more expensive than nuclear fuel produced from material here,[3] but the U.S. government encourages this remanufacturing process because it does not want the plutonium to fall into the hands of terrorists.

The H in H-Bombs: The
Tritium Shortage

Curiously, though, while the United States is experiencing a glut of nuclear fuel, it is experiencing a tritium shortage. Tritium is a radioactive form of hydrogen—the H in the H-bomb. The government is slowly running out of tritium because its stockpile decays at a rate of 5.5 percent a year. As well as being used in H-bombs, tritium is also used to increase the power of other atomic weapons, and President Clinton has signed an order to increase the ability to make more tritium by the year 2005. But the Atomic Energy Commission's (AEC) reactors for producing military grade nuclear products were last used in 1988, and are now in such disrepair that they are inoperable. Government plans to build a new weapons fuel reactor from scratch stopped when the planners realized that it would cost more than $9 billion.[4]

To save money, the Energy Department, a descendant of the Manhattan Project and the AEC, plan to load four special rods designed to produce tritium into the Tennessee Valley Authority's (TVA) Watts Bar reactor during its September 5, 1997 refueling. These special components are long, slim, stainless-steel rods similar to those containing boron that utilities use to damp nuclear reactors by sopping up excess neutrons. But the Energy Department rods are filled with lithium, so that when these rods are struck by neutrons, the lithium in them is split into helium and tritium.[5]

This experiment in tritium production will be done in exchange for a $7.5 million "irradiation fee" from the Energy Department. Stephen M. Sohinki, the Energy Department's project director, says that the refueling is not an experiment: "It is a confirmatory test to give confidence to the utility, the Nuclear Regulatory Commission (NRC) and the public that making tritium in a commercial light-water reactor is technically straightforward and safe."[6]

Blurring the Lines Between Weapons
and Electricity Production

Cost considerations are blurring the lines between the private production of nuclear-generated electricity and the production of weapons grade nuclear materials. Refuelings like the Watts Barr tritium irradiation exercise violate the tradition[7] that maintains a wall between private industry production of nuclear power and public agency production of weapons grade nuclear material. Still, the Energy Department has asked other civilian utilities to load tritium-producing rods into their reactors.[8]

Of 108 commercial reactors in the United States, only in the commercial light-water reactor at Watts Bar is it legal for the Energy Department to produce tritium. Under U.S. nuclear non-proliferation treaty obligations, the U.S. government cannot use a reactor that runs on uranium from another country or on uranium processed abroad for military purposes.[9]

"There was a belief for a long time [that] you had to keep the civilian and military sides absolutely separate, so the public wouldn't make the connection between power plants and bombs," said one nuclear power executive. "But that belief has dimmed," he said, "perhaps because no civilian power reactors have been ordered in this country for 20 years, and none will be any time soon."[10]

U.S. Government Contracts
Out Uranium Refinement

While the government is contracting out their production of tritium at the same time it is planning to sell the nation's largest plants producing military grade nuclear products.[11] These plants are run by USEC, the world's largest producer of nuclear power plant fuel. These plants supply 80 percent of U.S. nuclear power plant fuel. USEC owns two large, complex nuclear uranium enrichment plants, one near Paducah, Kentucky, and the other near Portsmouth, Ohio. The government is expected to get over $1.6 billion for these facilities in a deal that may be the largest sale of government property in history.

When the government sells the USEC plants, the leading producers of nuclear fuel in the U.S. (and the world) will pass into private hands. Aside from the physical facilities, the purchaser will own the rights to use the U.S. government's now-secret process for enriching uranium. This process is called the Atomic Vapor Laser Isotope Separation (AVLIS) process and it is a much faster way for collecting and concentrating the more volatile isotopes of uranium so that they can be used as either nuclear fuel or in nuclear weapons.[12] The most advanced commercial processes enrich uranium by pumping gaseous uranium through porous membranes to concentrate the fractions that can be used. The uranium is enriched to a 5 percent level for use in nuclear reactors and to a 93 percent level for use in nuclear weapons. The prospective purchaser would be able to use this process and it might dramatically reduce the cost of producing nuclear fuel, and thus nuclear generated electricity.

In the past, the aura of the nuclear bomb may have unrealistically enhanced the prospects for nuclear power generation, and the government may now be puffing their new enrichment process to get a good sale price for USEC's assets. The AVLIS process is now only in the prototype stage, but the United States intends to spend billions to develop it. Although it might prove to be a quicker, cheaper way to produce nuclear fuel, the technology may only be efficient when used to produce the highly enriched uranium that is used in bombs, rather than the lower grades used in nuclear reactors. A spokesman for GE has said "[The

General Electric Company] has absolutely no interest in that at all." But West-inghouse Electric Corporation said, "We are following it closely."[13]

The sale of these nuclear facilities is relevant to the second most important issue in nuclear power industry—how to reduce the cost of nuclear generated electricity. Unless this rather far-fetched technology can reduce these costs, most of the nuclear generating plants in the U.S. will probably be scrapped within ten years.

The Candu Process

Canada's nuclear authorities have an almost opposite approach to making cheap nuclear power available. Canada's Candu process, whose name is short for Canadian Deuterium-Uranium, uses unrefined uranium as fuel rather than the ex-pensive enriched uranium.[14] Using uranium as it comes out of the ground cuts the expensive enrichment stage out of the nuclear generation process entirely, making power production much less expensive, and potentially puts nuclear re-actors within the reach of poorer countries with less developed industrial infra-structures.

Canada had produced workhorse models of these reactors as part of a major national export initiative. The Candu reactors have been sold to Argentina, Ro-mania, and South Korea.[15] Although the high price of nuclear generation has dampened Candu sales in recent years, China recently ordered $3 billion of the reactors,[16] even if the Candu reactors can reduce the costs of nuclear generation enough to make it competitive with the latest technologies of coal and gas-fired generation, nuclear power reactors have an unsavory reputation. They are gener-ally viewed as unsafe, polluting, and difficult to manage and maintain.

In August 1997, Ontario announced that they were shutting down seven of its nine operating nuclear reactors.[17] The reactors being shut down were the workhorse versions of the Candu reactor that Canada exported. The president of Ontario Hydro Corporation, one of the largest facilities in North America, re-signed before the announcement that its facility on Lake Huron would close down three of its reactors and mothball another.[18] One the largest nuclear sites in the world, Hydro had played a major part in the Candu's development. The problems at Hydro were not with the Candu design, which was safe, but rather with the management of the reactors.

A History of Public Relations Disasters

Almost from the first, the nuclear generation industry in the United States has been plagued with public relations disasters. A popular movie dramatized the "China Syndrome" by graphically portraying a meltdown, and then Penn-sylvania's Three Mile Island suffered an accident suggestive of a meltdown.[19] Later, Chernobyl's actual meltdown provided the West with a much more hor-rible spectacle of spreading clouds of radioactive fallout. The disaster at Cherno-byl sickened and killed thousands of Russians, depending on their proximity to the meltdown and the wind currents prevailing at the time of the disaster. (Then the environmental movement decided to espouse the cause of nuclear safety.)

As a result of these scary events and public relations disasters, nuclear power came to be perceived as unsafe. In this environment at least one very expensive power plant, the Shoreham plant on Long Island, was built but never opened. The unsafe perception of nuclear power brought to a complete halt any investment in nuclear plants in the United States for the last twenty years.[20]

NUCLEAR POWER: TOO EXPENSIVE

Expensive to Build

Whether or not nuclear power is actually unsafe, in the last analysis, it simply turned out to be too expensive. The Long Island Lighting Company's (LILCO) Shoreham Plant is a leading example of an investment in nuclear generation that was too expensive. The "Shoreham debacle" is an 800-megawatt plant that was planned to cost $65 to $75 million, but ended up costing $6.5 billion.[21] It is true that safety issues played a part in the plant's demise, but the cost also included the additional expenses of construction delays, legal battles, excessive local taxes, and huge and escalating financial costs. Although the plant was completed in the mid-1980s and licensed in 1989, it will never be put into service because of political decisions based on public perceptions of nuclear power.[22] At best, the power plant would have produced electricity at about the national average of 3 cents per kilowatt hour had it been used. It is too expensive to even tear the plant down. The costs of the plant have made LILCO's electric rates the highest in the nation and have resulted in a state-sponsored refinancing of LILCO's debt and forced LILCO's merger with the smaller Brooklyn Union Gas Company (See Chapter 2 for details).

Expensive to Repair
Cracked Cooling Pipes

Adding to the expense of nuclear reactor maintenance is the need to replace old cooling lines which tend to develop cracks as they age. A cooling pipe that leaks radioactive liquid can render an entire reactor inoperable. Unfortunately, experience has shown that the metal in the miles and miles of these pipes tends to become hard and brittle before the end of its design lifetime because of its exposure to nuclear radiation.

In May 1997, Central Maine Power Company, a 38 percent owner of the Maine Yankee nuclear plant in Wiscasset, Maine, decided to cut $41 million from their 1997 maintenance budget of $190 million and lay off 900 employees. In 1998, Central Maine closed the plant permanently. "Economic concerns" were the reasons cited: the rising cost of safety measures were making their power too expensive to sell in a deregulated market that allowed fewer monopolies on customer bases.[23]

Dangerous and Expensive to Close Down

Northeast Utilities also has had particularly bad luck with their nuclear reactors. In March 1997, Northeast threatened bankruptcy if New Hampshire would not let them recover their stranded nuclear costs by raising their power rates.[24] By September, Northeast had been forced to close down all four of their reactors and had brought suit against the state of Connecticut for a reimbursement of $426 million for the state's share of their costs of decommissioning the thirty-year-old Connecticut Yankee nuclear plant.[25] The state claimed that they owed only half that much, but then, at that time, they only knew about half of the problems.

Unreported Spills

The surprise came, during the proceedings, when evidence disclosed that the plant had been widely contaminated by two separate leaks of nuclear fuel. The leaks had resulted in an astonishing amount of unreported nuclear contamination:

An expert in nuclear energy who was hired by the state reported that radioactive contamination had spread throughout the Connecticut Yankee plant in Haddam Neck and to the soil and asphalt outside, with evidence of radioactive particles recorded in the plant parking lot, its septic system, the silt of its discharge canal and water wells, and a shooting range on the plant site almost a mile from the reactor.[26]

Although the contamination was not thought to be dangerous, and was not thought to be unusual for an old reactor, Governor John G. Rowland said he was "frankly appalled" at the consultant's findings. He referred to the utility's management as "reckless."[27] The operators had referred to the discharge of radioactive fluids as "dumping," but had not reported them to regulators. After a 1989 leak of radioactive particles into the reactor's surrounding coolant, the plant operated for a record 461 days without doing anything about the spill. State officials called for a fence to be built around the facility and said that if the contamination was serious enough, tons of radioactive soil and even the entire plant structure might have to be hauled away.[28]

This situation shows how an attempt to repair or decommission a nuclear reactor can turn into the much larger job of decontaminating a nuclear waste dump. The state said that they should not be held responsible for many of the expenses involved in closing down the plant because there was evidence that Northeast Utilities had tried to cover up the nature of the problem with the plant.

Robocad Cleanup Technology

Once a reactor leak has occurred it is dangerous, and difficult to clean up by hand. The problem is so severe that it is engendering new technologies. But although U.S. problems with nuclear pollution have been serious and very expensive, they have not been nearly as serious as those suffered by Russia. The Chernobyl No. 4 meltdown in 1986 spewed out over 50 tons of volatile radioactive particles.[29] After killing or sickening from tens to hundreds of thousands of Russians, the clean-up problem is still not solved. Now under a more humane political regime, Russia no longer simply sends humans in to do the work by hand, almost assuring that they would die from radiation poisoning. This kind of work will probably be done by robots in the near future. Tecnomatix Technologies Ltd., an Israeli software company, has developed three-dimensional graphic-simulation software, know as *Robocad*, to plan and manage the robotic cleanup of Chernobyl and other bad nuclear spills.[30]

COMPARATIVE COSTS

Traditional Sources

If nuclear generation is the most expensive way to produce electricity, on a brighter note, gas-fired turbines are currently the most widely used emerging low-cost generation technology. A new generation of gas-fired electric plants can generate electricity for as little as 2 cents kilowatt hour.[31] Efficient coal-fired plants can generate electricity for as little as half of the current 3 to 3.5 cent cost of average gas-fired generation. "Nuclear operating costs," says Charles Komanoff, an energy economist, "exclusive of construction, are about the same as buying power from the grid, about 3 cents a kilowatt hour. Over all, [nuclear power] is a big fat wash."[32] Komanoff should have added that it is unusual in the United States for the extra costs of nuclear generation not to push the cost of nuclear power far above 3 cents a kilowatt hour.

Alternative Energy Costs

Wind Turbines

The cost of generating wind power has fallen sharply from about 12 cents a kilowatt-hour a decade ago to about 4 to 4.5 cents per kilowatt hour. This is still about 30 percent more expensive than gas-fired electrical generation. Current tax subsidies for wind generated power amount to about a penny per kilowatt hour over the long run.[33]

Solar Power

Solar power is still far too expensive to compete with conventional electric generation, although it is used on the Mir space station and is becoming more popular in the form of innovative new products. To be competitive with conventional coal and gas-fired generation, solar power equipment would have to cost from $1 to $1.50 per peak watt of capacity.[34] A peak watt is the amount produced when the sun is strong and directly overhead. A system, with storage batteries capable of storing about 5,000 peak watts, could run a typical suburban house with central air-conditioning.[35] There are not any authoritative price indexes for solar cells, but brisk demand for solar products has sent their prices up to approximately $4 per peak watt of capacity according to Thomas J. Vonderharr, the vice president of sales at one of the leading cell suppliers, BP Solar

Table 6.1
Uses of Solar Energy

Replacing 'Old' Energy	Sacramento, Calif., shut down its Rancho Seco nuclear power plant in 1989. Today, 1,750 solar panels are on the site supplying power to 700 homes.
Solar Powered	United States solar cell manufacturers want to increase the use of solar energy in the domestic market. They hope to expand these existing uses of solar cells.
Road Signs	Warn motorists of construction projects.
Carport	At the Army Proving Ground in Yuma, Arizona, solar energy supplies the power to light a parking lot. It also serves as a charging station for electric vehicles.
Sailboats	Solar panels take the place of a diesel engine providing a quieter ride for hours at a time.
Building Designs	Solar panels are being used as building materials in a 48-story office tower under construction on 42nd Street in Manhattan The panels will generate power equal to that used by seven houses in a year.

Source: Matthew L. Wald, "An Industry Relishes Its Day in the Sun," *New York Times*, August 16, 1997, pp. 35-36.

Inc., of Fairfield, California.[36] Some of the solar generation products are shown in Table 6.1.

Solar power "output would barely register as more than a blip in the huge American energy market."[37] Solar power is currently used in the United States mostly as residential solar equipment whose generation is meant to fill in gaps not adequately served by electric power off the national power grid. Such marginal use of means that it is almost overlooked in the United States although the United States has a technological lead in manufacturing solar cells, exporting about 70 percent of total solar cell production.[38]

Limiting Nuclear Exposure

Enron Corporation is one large utility that seems to have gotten both ends of the cost equation correct, for Enron has limited its exposure to high-cost nuclear generating plants by specializing in gas-to-electricity convergence. Enron's 1997 merger with Portland General Corporation pulled together a large Oregon electrical power company and their own, non-contiguous, Texas-based business in natural gas. Although at first this seemed like an odd pairing, it simultaneously gave the merged firms massive access to gas fuel for their modern, low-cost gas-fired generation plants, as well as access to Portland's electricity marketing skills and customers. In fact, what had attracted Enron to Portland General was their low-cost generation of electricity and their low exposure to nuclear powered projects.[39]

It is no accident that Enron is currently selling gas and electricity in New Hampshire and New York. These two states are among the first to open their residential power markets to out-of-state competitors: they are also states that have utilities that are hamstrung with huge amounts of stranded nuclear costs. These high stranded costs make New York utilities vulnerable to out-of-state competition, but New York authorities are determined to deregulate their electrical utilities and make these utilities compete. To this end, the state has already allowed Texas-based Enron to book energy orders from a number of large New York industrial firms.

MANAGERIAL PROBLEMS

Managers and nuclear consultants almost always explain nuclear plant problems as the result of poor management. This includes problems with unprofitable nuclear operations, poorly run nuclear plants, and even nuclear meltdowns. To some extent, "poor management" is merely the standard rationalization of managers and consultants that are sympathetic to the use of nuclear power. But the facts indicate that most problems *are* related to managerial sloppiness, and you cannot be sloppy when you're managing a nuclear power plant. Some observers have suggested that "good management" of nuclear facilities must be done by a "nuclear priesthood." Managers with such a dedication might be necessary because of the peculiar risks of nuclear reactors: nuclear reactors have complex sets of failure modes and the plutonium byproducts are extremely toxic.

Because nuclear reactors produce heat to run turbines, cooling systems are important. But nuclear cooling systems also seem to have a peculiar set of risks. Cooling systems are critically necessary filled with radioactive materials, and are difficult to repair and maintain because they are radioactive. The history of nuclear power plants has proven that combinations of these particular problems are truly difficult to operate and manage.

The 1997 failure of Ontario Hydro's Candu reactors was typical. "Our problems are people," said William A. Farlinger, chairman of Hydro Ontario, "not machines. The nuclear unit was operated over all those early years as some sort of a special nuclear cult. Senior management didn't dig into what was going on in this special unit to the extent that we must now say they should have."[40] According to the investigation, the Candu operators had inadequately maintained their facilities and had not replaced parts that had incrementally eroded away the margin of safety which was supposed to be built into them.[41] The problem with replacing a nuclear component *after* failure is that the part and its environment may be radioactive. Industrial robots were necessary to mop up at Three Mile Island after that reactor failure. For this reason safety is especially important in nuclear facilities. In the case of Ontario Hydro, one of the largest utilities in North America, "an internal report concluded that the province's utility company was so badly managed that it had compromised the safety of its entire nuclear power system."[42]

As for the dozen reactors that Ontario Hydro will continue to run, it will cost $1.5 billion between now and the year 2001 to overhaul them and maintain their proper safety specifications.[43] The shuttered reactors will be brought back on line only if it is cost effective to do so. The problems sound similar to those on Russia's Mir space habitat.[44]If components are not replaced at the end of their service lifetimes, it is often more risky and expensive to replace them after a random catastrophic failure. "We used to change Mir's computer parts after their technical life expectancy ran out," said Victor Blagov, the deputy flight chief, the day after the Mir's docking computer had failed during the supply module's rendezvous. "Now, because of problems with money, we must use each part until it dies."[45]

Like the Mir spacecraft, nuclear reactors also have surprising and complex failure modes, and after a mishap the site of the problem is often isolated from ready repair. A nuclear reactor creates heat through the emission of neutrons from rods of enriched uranium in a controlled chain reaction. A system of boron damping rods absorbs the neutrons to control the speed and heat of this reaction. Although this sounds straightforward, the specifications of the cooling system, the damping system, and other systems that control the reactor often have complex interactions that are difficult to model. Although the sets of events involved in meltdowns and accidental releases of radioactive gases seem obvious in retrospect, it is difficult to predict in advance what the really dangerous congruences of events will be. This means that it is difficult to predict, train, and plan for highly effective nuclear reactor operations.

Economies of Scale of Management

A lot of U.S. utilities have merged to achieve the size necessary to accomplish basic business functions in the large, competitive U.S. power market. Individual utilities—which are often worth billions in dollar terms—found that they were not large enough to properly advertise themselves, to put together attractive product lines, or even to manage their nuclear reactors. Although worth billions, they were still so small that they lacked either the requisite expertise or capital to do the job right. Some examples of things that large individual utilities have not been able to accomplish alone are:

- National advertising programs (UtiliCorp-Peco)[46]
- One-stop power marketing of gas and electricity (Enron-Portland Gas)[47]
- One-stop power marketing of energy-related products (UtiliCorp-Peco's EnergyOne program)[48]
- Management of nuclear reactors (Shoreham and Maine Yankee)[49]

The U.S. energy market has already become massive and competitive enough that mergers to achieve economies of scale of management and financing are necessary to get basic jobs done, and this may be necessary to properly manage nuclear plants.

When is a utility big enough and smart enough to manage nuclear power projects? The point is surprisingly subtle. Billion-dollar utilities may have thought themselves large and skillful when they embarked on the construction and management of nuclear generation facilities after World War II. But nuclear reactors have proven to be tricky, complex, problematic, lethal, and expensive money losers. In retrospect it is reasonable to say that utilities were not large enough to be smart enough—or skillful or experienced or clever enough—to manage their own nuclear reactors, because they repeatedly made the same kinds of managerial mistakes. Management economies of scale are essential to the successful management of nuclear generating reactors, and many utilities should either spin off their reactors, merge to assemble the necessary managerial skills, or contract out those skills from a larger, more expert utility, such as Entergy.

Contracting Out Plant Management

Entergy's Specialists Sell Their Skills

Entergy Incorporated, a New Orleans-based utility, was the first utility to move aggressively into the business of running nuclear power plants for other electricity companies. As competition increased the pressure on utilities to reduce costs, Entergy began selling its knowledge of nuclear power plant management to other, less efficient nuclear operations.

In January 1997, Entergy agreed to manage the troubled Maine Yankee nuclear plant.[50] Maine Yankee's 850 megawatt nuclear plant supplied 20 percent of the electricity of the state of Maine. The plant created 700 jobs and was the leading employer in Wiscassett, forty miles northeast of Portland. The plant was a low-cost producer with a good performance record until 1995. But in the Summer of 1996 an inspection by the Nuclear Regulatory Commission showed that although its safety was adequate, the safety of its future operations was in question. The plant's problems intensified in recent years, and led to costly shutdowns and repairs as opponents of nuclear power campaigned to close the plant.

Entergy's subsidiary, Entergy Operations Inc., runs five nuclear plants for its parent, and is actively soliciting more business like that in Maine. Entergy Operations have trimmed up the efficiency of its own nuclear plants and this track record helped get them the contract to manage the Maine Yankee plant for the Central Maine Power Company.

Entergy's management contract with Central Maine linked its compensation to how successful it was in improving the Maine Yankee plant's performance. Other utilities like Peco Energy of Philadelphia, are also starting to manage nuclear plants, so if contracting out the management of nuclear plants becomes a trend, Entergy's contract may become a model for the industry. Although the Maine Yankee plant was buying costlier power than would eventually produce, Entergy planned for the plant to be back in operation by mid-February 1988. Entergy did not plan to reduce the work force at Maine Yankee. After implementing their improvements, Entergy expected that the plant would operate to the end of is licensed life in 2008. Unfortunately, the plant's costs overwhelmed it and the plant was closed in 1998.

Entergy's nuclear plant management contracts guarantee results or they will not get paid. This approach denies that the size of the utility is the problem, but rather asserts that you can gain the necessary management skills by simply concentrating them in one small place—a subsidiary like Entergy Operations —and then apply these skills by selling them. Certain managers may be smarter than others and perhaps firm size has nothing to do with competence of management. Of course Entergy is a huge power-marketing utility, and maybe this has something to do with their success in collecting good teams to manage nuclear plants.

Other corporations which have shown a willingness to contract out or contract in expertise have been Louisville Gas & Electric (LG&E) and Oglethorpe Power and Light Company.[51] Both LG&E and Oglethorpe contract out low-cost coal-fired electricity and marketing skills. LG&E is also negotiating long-term power contracts with Enron and Duke/Dreyfuss,[52] and Enron has sold energy management skills to the Northern California Power Authority.

Unfortunately, even Texas management consulting was not enough to save New England utilities from their nuclear plant problems and stranded costs. In March 1997, Northeast Utilities asked the federal district court in Concord, New Hampshire, to restrain the Public Utilities Commission's ruling that limited their ability to pass on the stranded costs of their nuclear generating plants to their customers in the form of rate increases.[53] Northeast said the Commission's ruling would cut their cash flows by $340 million over two years and threaten them with bankruptcy. In May 1997, led by Central Maine Power Company, which has a 38 percent interest in the Maine Yankee plant, the owners said that they were cutting their maintenance spending on the plant by 40 percent but it was necessary to close the facility in 1998.[54]

Consolidation and Expansion

Sellers

The New England Electric System is one utility that, discouraged by the stranded costs and high costs of generation, has given up on the power plant end of the business and decided to concentrate on selling and transmitting power.[55] In February 1997, New England Electric put up eighteen of its plants for sale.[56] Boston Edison Company also auctioned off $500 million of plants in 1998. And although Pacific Gas & Electric Company (California) has $1.1 billion of plant assets on the block, the sale of these plants was a forced divestment unrelated to nuclear power costs.

Buyers

New England Electric's assets were expected to attract bids of approximately $1.1 billion. CalEnergy Company (California), Duke Power Company (North Carolina), and Southern Company (Atlanta) all considered bids on the plants. But in August 1997, Pacific Gas & Electric Corporation (PG&E), the largest utility in the United States, bought the fossil fuel and hydroelectric plants for $1.59 billion.[57] The sale signaled the exit of New England Electric from the power generation industry, and left PG&E competing to supply power in Massachusetts, New Hampshire, Rhode Island, and Connecticut. Although PG&E entered these markets through their purchase of New England's assets, the electricity that PG&E supplies might come from transcontinental purchases made any-where in the United States. New England, the second largest utility in the region, said that the sale would lower their average electric rates by 15 to 18 percent.[58] This purchase and sale of assets reinforced a trend of utilities to specialize in ei-ther power generation or power transmission and sales.

Although the New England area is famous for high cost, inoperable nuclear power plants, the region's assets that have attracted the interest of buyers are nonpolluting hydroelectric plants and modernized natural gas plants. None of the plants sold by New England Electric are nuclear generation plants, because nuc-lear plants are unmarketable. The traditional technology plants, however, are val-uable because they are reasonably efficient and because they are located in states with high per capita incomes with high legal barriers to entry of the utilities in-dustry. But the plants don't come with any long-term power contracts. The buy-ers are expected to sell in competitive power markets.

If New England's utilities sell or abandon their high-cost nuclear generating plants, spin off their sales and transmission assets, and go bankrupt against their stranded costs, the ratepayers would win, but the utility's bondholders would most likely suffer.

THE STRANDED COSTS OF NUCLEAR POWER PRODUCTION

The legacy of overinvestment and the public's opinion that nuclear power is unsafe lives on in the stranded costs of nuclear power. Most of the nuclear pow-er plants in the United States were financed with debt issued by utilities that were state monopolies. The bond issues representing that debt were sanctioned by public utilities commissions that monitored the rates and financing of the utilities. These commissions allowed utilities to charge rates high enough to guarantee that the utilities would be able to pay the interest on the utility's bonds. The proceeds from the bonds funded the utility's capital investments in its power plants. The fact that the bonds were backed by low risk cash flows of monopolist utilities made the bonds safe. Thus the utilities were able to sell bonds with low risk and low yields and got cheap debt financing for the multibillion dollar in-vestments necessary to provide power for growing local economies.

Now all of this has changed. Public power commissions are being phased out as the supply of U.S. power is deregulated under the leadership of Federal Ener-gy Regulatory Commission (FERC). Utilities are rapidly merging to form larger units to take on competition in a national rather than a regional market for ener-gy. And it is publicly owned companies that must pay for the billions of dollars of stranded investment in marginally competitive nuclear generation plants. The decision of whether stockholders or taxpayers (and which taxpayers) will bear the burden of the stranded costs is cloaked by politicians in complex plans to merge public utilities and refinance their investments. Driving all these plans is the need to make sure all utilities, even those with massive stranded costs, are able to charge low electric rates in a competitive power market.

Table 6.2
Stranded Costs of Utilities

Stranded Costs	$ Billions	Percent
Nuclear Plants	$86	43%
Purchase Energy Contracts	$51	25%
Other Stranded Costs	$65	32%
Total Stranded Costs	$202	100%

Source: Jeff Bailey, "Niagara Mohawk Plan is a Small Step Toward Easing Utilities' Power Woes,"
    Wall Street Journal, March 12, 1997, pp. A1, A4.

The Magnitude of the Problem

Although estimates vary dramatically, the total amount of stranded costs in U.S. utilities is agreed to be from $45 to $400 billion, with almost 40 percent ($86 billion) of this accounted for by nuclear costs.[59] Estimates of the value of stranded costs varies a lot. A breakdown of the sources of stranded costs is shown in Table 6.2.

The problem is important, because the nuclear generating plants cannot simply be abandoned. Some areas rely heavily on nuclear power, and the more than 100 nuclear plants operating generate nearly 25 percent of the nation's electricity. This means that all parts of the country share, in some degree, the problem of stranded nuclear costs.[60] Vermont, Connecticut, Chicago, New Jersey, and South Carolina have placed a particularly heavy reliance on nuclear generation, and all of these states produce at least 66 percent of their power with nuclear plants.

The policy issues involved with electricity prices and nuclear plant retirement are now being discussed in terms of how fast the write-off of nuclear facilities will proceed. The utilities are arguing for a controlled transition so they can evolve into a more efficient, market-competitive mode. This transition could take a decade, and many utilities would be hobbled or destroyed by stockholder fears of bankruptcy if the transition is too swift. Table 6.3 shows that the areas that would benefit the most from policies of gradual transition to competitiveness would be the ones with the heaviest reliance on nuclear power production.

The situation of weak utilities with stranded nuclear power costs and high debt as opposed to strong utilities with low cost, non-nuclear production is reminiscent of the merger situation in banking during the 1980s. Will FERC, like the Federal Reserve Bank, encourage the merger of weak nuclear producers with strong non-nuclear producers? Almost certainly not, because it would produce weak, high-cost utilities and distort the results of market competition.

The political actors pushing for rapid resolution of nuclear power problems are consumer groups and some non-nuclear utilities. Arguing for gradual transition are the utilities with heavy nuclear investments. There is very little reservoir of good will for nuclear plants, but adopting a strategy of putting them out of business fails to address key issues and problems.

Table 6.3
Areas Heavily Dependent on Nuclear Power Production

State	Nuclear Power as a Percent of Nuclear Electricity Production
Vermont	80
Connecticut	70
Chicago	70
New Jersey	66
South Carolina	66
Illinois	NA

Source: Barnaby J. Feder, "The Nuclear Power Puzzle. Who Will Pay for a Generation of Expensive Plants?" *New York Times*, January 3, 1997, pp. D1, D3.

Chicago's Well-Managed Plants

Although some areas with heavy reliance on nuclear generation, like New England, appear to be abandoning power generation along with the stranded costs of their nuclear plants, the Commonwealth Edison Company in Chicago is a well managed utility that supplies 70 percent of the Chicago area's electricity from 12 nuclear plants.[61]

Commonwealth Edison's nuclear plants are efficient enough to compete in deregulated markets. Although some of Edison's nuclear plants are among the most inefficient (with accumulated debts of as much as $10 billion), and they are incurring huge renovation expenses, Edison welcomes competition and asks only that they be given a reasonable period to manage the transition to deregulated markets. "If we can't figure out a way to make nuclear power competitive with a competitive advantage, no one can," said Thomas J. Maiman, Edison's senior vice president for nuclear operations.[62]

Northeast's Stranded Costs

LILCO's problem with stranded costs in its Shoreham nuclear power plant was sufficient to push the company into a merger with Brooklyn Union Gas Company. Although extreme, LILCO's problems are symptomatic of similar problems across the United States.[63] Utilities with stranded costs in nuclear plants will become worse once FERC removes electrical power production from its regulatory cocoon. Every region of the United States has a stake in the problem. Problems like these have led Northeast Utilities (Hartford) to shut down its troubled Connecticut Yankee plant fourteen years ahead of schedule. Such shutdowns entail layoffs, job losses, huge tax loses, air pollution from alternative power sources, reduced capacity to meet electricity demands, and thus, higher electricity costs. Nor will the shutdowns solve the problem of how to pay the $70 bilion in debt used to finance U.S. nuclear plants, or provide the additional billions needed to pay for plant retirements.

Stranded costs posed such a serious problem for Northeast Utilities, that they found it necessary to go to court in March 1997 to block a state deregulation decision that would sharply limit the utility's practice of passing on costs of past investments to customers.[64] Northeast said that the ruling by the New Hampshire Public Utilities Commission would slash the company's revenue by $341 million over two years and might ultimately drive it into bankruptcy.

Northeast had attempted to stop the ruling by seeking a temporary restraining order from the federal district court In Concord.[65] The stranded costs of New Hampshire's utilities were mostly associated with high-cost nuclear power facilities which had not proven capable of supplying competitively priced power. The ruling by the commission was noteworthy because it was the first time that a state regulatory agency had attempted to prevent a utility from recovering their stranded costs by passing them on to ratepayers.

Steven Fetter, an analyst for Fitch Research, thought that the commission was taking a harsh position by supporting the view that the utility cannot get their customers to pay for the costs of their previous investments.[66] Northeast Utilities is a Massachusetts power utility holding company with 1.7 million electricity customers in three states. Providing these customers with electricity is an expensive proposition and Fetter thought that despite its size, companies such as Northeast Utilities would not be viable businesses unless they were allowed to recoup their costs through rate increases. The legal move by Northeast to protect itself is likely to be a foretaste of conflicts to occur between utilities and their regulators as the pace of electrical power deregulation picks up speed in the United States.

Suffolk County, Long Island, Imports Electricity

Although power may be uncompetitive and expensive in New Hampshire and Conncecticut, it is still cheaper than electricity on Long Island, New York, which has even higher stranded costs. In January 1997, FERC accepted Suffolk County's bid to import electricity.[67] Acting under provisions of the Energy Policy Act of 1992 designed to promote competition in electricity production, FERC approved the Suffolk County government's request to buy a significant portion of its power from Connecticut utilities, and sell it to customers through LILCO's transmission lines.

LILCO stated that the county's action would not result in lower electricity bills for its residents, because LILCO is allowed to charge fees for the use of their transmission lines. LILCO then appealed the FERC decision allowing the outside purchase. LILCO's customers pay the highest electricity rates in the U.S., largely because of $5.5 billion in stranded costs represented by the Shoreham nuclear power facility which was never allowed to operate. Suffolk County's actions were a test to see if the Energy Policy Act's provisions would help them avoid these stranded costs.

FERC's ruling was expected to have no effect on the pending merger between LILCO and Brooklyn Union Gas Company, or with LILCO's talks with the state about taking over their $4.5 billion in debt resulting from the abandoned Shoreham nuclear power plant.

Table 6.4
Approaches for Nuclear Reform

Approach	Advocates	Advantages
Quick transition to non-nuclear electricity	Consumer groups Some non-nuclear utilities	Decrease in rates. Let investors pay. End to the disgust with nuclear production's poor record, and fear of radiation and more high costs.
Gradual transition	Nuclear utilities: Connecticut New Jersey South Carolina Vermont Chicago	Preserve nuclear utilities (not necessarily an advantage). Give the utilities a fair chance to make a transition to better management and low cost, competitive production.
Middle course	Utilities and power commissions	Supports the attitude that although there is very little reservoir of good will for nuclear plants, putting nuclear plants out of business won't do anyone any good.

Sources: Jeff Bailey, "Niagara Mohawk Plan is a Small Step Toward Easing Utilities' Power Woes," *Wall Street Journal*, March 12, 1997, pp. A1, A4, and Barnaby J. Feder, "The Nuclear Power Puzzle. Who Will Pay for a Generation of Expensive Plants?" *New York Times*, January 3, 1997, pp. D1, D3.

SOLUTIONS

With 100 plants and 25 percent of U.S. power production accounted for by inefficient nuclear reactors, practical solutions must be found for the problem of stranded costs.[68] Some way must be found to pay for the $86 billion of stranded nuclear costs in a power industry that has invested about $202 billion in nuclear electricity generation facilities.[69] Table 6.4 presents roughly three courses of action that might be taken.

Anything that looks like state subsidization of nuclear power is highly unlikely to survive the current wave of free-marketeering deregulation. Agreements to subsidize stranded nuclear costs like Con Edison's programs to subsidize electricity for low-income customers will probably be cut. This means that purchase power agreements like those under which Con Edison and Mohawk buy power from independent power generators will be out. Both small generators and nuclear generation are too expensive to survive competition. Alternative energy programs will be quickly nudged into the competitive market rather than continue to receive subsidies for production that is often as much as 20 percent over the market price for electricity.

Table 6.5

Practical Solutions for Stranded Nuclear Power Costs

California	Has already adopted a variety of rules that give utilities several years to recoup their investments from consumers. Typically, they have softened the sting with related measures that result in freezes or even reductions in overall rates. A law allowing deregulation was passed in September 1997. A special charge of about one-third of an average monthly bill will be levied to pay for $29 billion in power plant debt.
Commonwealth Edison (Chicago)	If management is good enough, some plants will be efficient enough to sell on the market. Still, some of Edison's plants are among among the most inefficient in the U.S. Despite generally good management, these plants face huge renovation expenses. Edison has accumulated debts of $10 billion. Edison welcomes competition, but feels that it deserves time and financial breaks from regulators to prepare for it.
Connecticut	Three state reports have urged deregulation. To pay for existing power plants, a charge on monthly bills has been recommended for consumers who switch providers.
Massachusetts	Most of the state's utilities have agreed to accept competition by 1998. Utilities will add a surcharge of 3 to 3.5 cents per kilowatt hour to pay for $12.5 billion in debt for existing plants.
New Jersey	The utility control board wants competition in the wholesale market. It has not addressed how to pay for power plant debt.
New York	The Public Service Commission wants to deregulate by 1998. Last October, it asked utilities to submit recommendations.
Pennsylvania	A new law allows deregulation to be phased in by 2001. Utilities can add a surcharge to customers' bills for existing power plant debt, but the amount will be decided case by case.
Rhode Island	The law that deregulates utilities went into effect in July 1997. Full deregulation will occur by July 1998. To pay for existing plant debt, a surcharge of 2.8 cents per kilowatt hour will be levied the first three years, dropping gradually to 0.5 cents after 12 years and ending in 25 years.

Sources: Barnaby J. Feder, "The Nuclear Power Puzzle; Who Will Pay for a Generation of Expensive Plants?" *New York Times*, January 3, 1997, pp, D1, D3, and Jeff Bailey, "Niagara Mohawk Plan is a Small Step Toward Easing Utilities' Power Woes," *Wall Street Journal*, March 12, 1997, pp. A1, A4.

The prevailing feeling is that investors, consumers, and commissions agreed to past nuclear investments, and that they should pay for these plants. This means continuing high rates for the communities supplied by nuclear generation. But if the United States is to move to a competitive power system these emotional arguments are simply not practical. High-cost nuclear power can not compete unless its costs somehow come down so local companies can survive as competitive entities, otherwise the power company will go bankrupt. Some of the practical solutions for making nuclear power competitive or decommissioning it are shown in Table 6.5.

NOTES

1. Judith B.Sack and Robert L. Chewning, "Global Electricity Strategy: My Two Cents Worth (or the Sustainable Price of Power)," Morgan Stanley research report, International Investment Research, February 6, 1997.

2. John J. Fialka, "U.S. Plans to Sell Uranium Enrichment Operations," *Wall Street Journal*, August 14, 1997, pp. A1, A14.

3. *Ibid.*

4. Matthew L. Wald, "U.S. to Put a Civilian Reactor to Military Use," *New York Times*, August 5,1997, pp. A1, A20.

5. *Ibid.*

6. *Ibid.*

7. Although it is illegal to produce enriched uranium or plutonium, it is not illegal to buy or produce tritium, because you cannot make a bomb from it.

8. Wald, "U.S. to Put a Civilian Reactor," pp. A1, A20.

9. *Ibid.*

10. *Ibid.*

11. Fialka, "U.S. Plans to Sell Uranium Enrichment," pp. A1, A14.

12. *Ibid.*

13. *Ibid.*

14. Anthony DePalma, "Ontario to Shut 7 Border Nuclear Reactors," *New York Times*, August 14, 1997, pp. D1, A6.

15. *Ibid.*

16. *Ibid.*

17. *Ibid.*

18. *Ibid.*

19. Perez-Pena, "Lilco's Hard Journey; Road to a State Takeover Began with Debacle of the Shoreham Plant," *New York Times*, July 21, 1997, p. B4.

20. Wald, "U.S. to Put a Civilian Reactor," pp. A1, A20.

21. Bruce Lambert, "Hot Issue in Lilco Takeover Talks: Who Pays for Shoreham?" *New York Times*, March 16, 1996, pp. 41-42.

22. *Ibid.*

23. Ross Kerber, "Maine Yankee Power Plant's Owners Say Facility May Be Shut Permanently," *Wall Street Journal*, May 28, 1997, p. B4.

24. Agis Salpukas, "Northeast Utilities Sues to Block Move by New Hampshire," *New York Times*, March 4, 1997, p. D8.

25. Jonathan Rabinovitz, "Hartford Says Utility Hid Nuclear Contamination," *New York Times*, September 16, 1997, pp. B1, B6.

26. *Ibid.*

27. *Ibid.*

28. *Ibid.*

29. Marina Lakhman, "A Virtual Cleanup of Chernobyl; Israeli Software Markets Simulation Using Robots," *New York Times*, September 1, 1997, p. D3.

30. *Ibid.*

31. Sack and Chewning, "Global Electricity Strategy."

32. *Ibid.*

33. Allen R. Myerson, "Enron Wins Pact to Supply Power from Wind Turbines," *New York Times*, May 27, 1997, p. A5.

34. Matthew L. Wald, "An Industry Relishes Its Day in the Sun," *New York Times*, August 16, 1997, pp. 35-36.

35. A peak watt is the amount produced when the sun is strong and directly overhead. A system, with storage batteries, capable of storing about 5,000 peak watts could run a typical suburban house with central air conditioning.

36. Wald, "An Industry Relishes," pp. 35-36.

37. *Ibid.*

38. *Ibid.*

39. Terzah Ewing, "Enron and Portland General Reduce Stock-Swap Ration and Boost Rate Cuts", *Wall Street Journal*, April 15,1997, p. A4.

40. DePalma, "Ontario to Shut 7 Power Plants," pp. D1, A6.

41. *Ibid.*

42. *Ibid.*

43. *Ibid.*

44. Interview with Russian astronaut Vasily Tsibliyev, astronaut, interviewed upon return in Michael Specter, "Russian Astronauts Insist Errors, While Human, Were on Earth," *New York Times*, August 17, 1997, pp. 1,8.

45. Michael Specter, "Jeers Sting Mir Mission Control, Which Bemoans a Money Pinch," *New York Times*, August 20, 1997, p. 1.

46. Benjamin A. Holden, "UtiliCorp and Peco, Aided by AT&T, to Launch One-Stop Utility Service," *Wall Street Journal*, July 24, 1997, pp. A1, A3.

47. Benjamin A. Holden, "Enron Corp. Has Accord to Buy Portland General," *Wall Street Journal*, July 22, 1996, p. A3; Sullivan Allanna, "Enron Deal Signals Trend in Utilities," *Wall Street Journal*, July 23, 1996, p. A3; and Allen R. Myerson, "Enron Will Buy Oregon Utility in Deal Valued at $2.1 Billion," *New York Times*, July 23, 1996, p. D1.

48. "Home-Services Alliance Forms," *New York Times*, June 25, 1997, D1, D5.

49. Kerber, "Maine Yankee Power Plant's Owners," p. B4.

50. Agis Salpukas, "Entergy Agrees to Manage Troubled Maine Nuclear Plant," *New York Times*, January 9, 1997, pp. D1, D6, and and Susan Jackson, "The Millstones Around NU's Neck," *Business Week*, December 2, 1996, p. 80.

51. Agis Salpukas, "Utility Seeks Partner to Aid Power Needs," *New York Times*, November 20, 1996, pp. D1, D2.

52. Salpukas, "Utility Seeks Partner," pp. D1, D2.

53. Agis Salpukas, "Northeast Utilities Sues to Block Move by New Hampshire," *New York Times*, March 4, 1997, p. D8.

54. Kerber, "Maine Yankee Power Plant's Owners," p. B4.

55. Ross Kerber, "Auction of 18 Power Plants is Igniting Utilities' Interest," *Wall Street Journal*, February 6, 1997, pp. A1, B4.

56. Kerber, "Auction of 18 Power Plants," pp. A1, B4.

57. Charles V. Bagli, "PG&E Will Buy 18 Power Plants in New England," *New York Times*, August 7, 1997, pp. D1, D2.

58. *Ibid.*

59. Barnaby J. Feder, "The Nuclear Power Puzzle. Who Will Pay for a Generation of Expensive Plants?" *New York Times*, January 3, 1997, pp. D1, D3.

60. *Ibid.*

61. *Ibid.*

62. *Ibid.*

63. *Ibid.*

64. Salpukas, "Northeast Utilities Sues," p. D8.

65. *Ibid.*

66. *Ibid.*

67. David Stout, "U.S. Accepts Suffolk's Bid to Import Electricity," *New York Times*, January 3, 1997, p. B2.

68. Barnaby J. Feder, "The Nuclear Power Puzzle; Who Will Pay for a Generation of Expensive Plants?" *New York Times*, January 3, 1997, pp. D1, D3.

69. Jeff Bailey, "Niagara Mohawk Plan is a Small Step Toward Easing Utilities' Power Woes," *Wall Street Journal*, March 12, 1997, pp. A1, A4.

Chapter 7

Winners and Losers

WINNERS

Large Industrial Customers

Large industrial customers are sure winners from the price-cutting that comes from competition. Many large firms have already benefited by cutting deals for cheaper power.[1] Small suppliers like Brooklyn Power now have to worry about competition with nationwide giants like Enron and PanEnergy Corporation. Enron now has 22,000 customers and expects to add another 11,000 in 1997.[2] New moves to open markets, such as those expected in New York, will give them even more market power.

Mergers and Acquisitions Specialists

Wall Street's mergers and acquisitions specialists are getting rich off of the surge in mergers as the fragmented utilities industry continues to consolidate. The magnitude of the fragmentation is illustrated in Massachusetts, where there are currently sixteen gas and electric utilities, but where, with mergers, the market would be too small for more than two utilities.[3] Aside from their expense, another problem with mergers may be that utilities have become overcapitalized through structural faults in the way that U.S. laws allow utility holding companies to merge.[4]

Jon Kingstad points out that there is no effective U.S. law limiting the financial actions of utility holding companies. This means that Wall Street is able to pressure board members of these holding companies into issuing excessive amounts of debt to arrange mergers that do not benefit their rate payers. The pressure is applied with a carrot and stick. The carrots are the liberal shares in the merged corporation that board members of the holding companies often pay to themselves when they vote themselves out of power during the merger. The stick is the guarantee of state regulatory approval under the implicit threat not to finance future power projects in a state if commissioners do not approve the merger. According to Kingstad, this situation tends to result in the overcapitalization of merging utilities through the overuse of debt financing. The excessive debt of the merged utilities makes them vulnerable to illiquidity crises because they have become saddled with coupon payments that are too high for their fluctuating cash flows to handle.

The stick, but not the carrot, is used on the commissioners of state utility oversight boards. These boards are very anxious to keep open their utilities' channels of future financing and Kingstad claims that they will approve deals with too much debt and too little rate cuts in order to do this. What Kingstad says is obviously true in many mergers, but this criticism seems a narrow one when compared to the breadth of the deregulation and competition goals of FERC. Kingstad argues, however, that to push back effectively against this kind of financial pressure, a law is needed to give FERC the leverage to support local commissions in the case of bad mergers and bad financing.

CalEnergy Bids for NYSEG

CalEnergy's unsuccessful bid for New York state Electric and Gas Corporation (NYSEG) illustrates how wasteful some takeover attempts can be. When CalEnergy dropped its takeover bid for NYSEG, they said that NYSEG had "engaged in a scorched earth campaign which, in the span of just a few weeks, wasted an enormous amount of money that could have been used to create value for shareholders and lower rates for customers."[5] A federal judge overseeing litigation between CalEnergy and its takeover target criticized the NYSEG board for "frivolous and unrelenting litigation efforts and a misinformation campaign" that had been effective in convincing NYSEG shareholders not to accept CalEnergy's offer.[6] "The only beneficiaries in this process appear to be NYSEG's board and investment bankers and lawyers, who will split over $20 million in fees,"[7] CalEnergy said.

Differences of Opinion

Sometimes utility merger fights are driven by strategic differences in opinion. In these cases, even if the takeover is completed, it is often difficult to tell if there is a winner. This was the case in March 1997, when Western Resources Inc. made a $4.3 billion "low ball" bid for ADT, a large manufacturer of home security devices. The issue was whether or not ADT's security products would fit in with Western Resources strategy of one-stop energy marketing. The dispute became public when Tyco International made a friendly $5.4 billion bid that topped Western's initial hostile offer. Tyco thought that their fire protection products fit better with ADT's home security products than electricity did. Although Western Resources had purchased Westinghouse Security Systems in December

1997 for $368 million, they found themselves reassessing the value of ADT before deciding whether to make a responsive bid for ADT.

Hubris Among Giants

Adding fuel to the takeover fire is the pride of the personalities of the managers driving the merger efforts. Finance Professor Richard Roll is famous for asserting that one reason for mergers is the "hubris" motivating managers, so this theory has the impramature of financial academics.[8]

Western Resources once again provides a case in point. In February 1997, Western completed a nine-month hostile takeover of Kansas City Power & Light Company (KCP&L) for $2 billion.[9] John E. Hayes Jr., the chairman and chief executive of Western, had led the bitter battle to wrest KCP&L away from UtiliCorp, a Kansas City-based natural gas company and KCP&L's friendly merger partner. At one point during the merger struggle, A. Drue Jennings, the CEO of KCP&L accused Hayes and his advisers of waging an "increasingly nasty and personal campaign against KCP&L's board and management and especially me."[10] Although both executives had made efforts to defuse their personality conflicts, their legacy of bitterness was expected to complicate the merger.[11]

Some merger observers feel that hubris and personal antagonism are a natural part of the merger process. They feel that although these personal enmities may stimulate vigorous litigation and huge attorneys fees, the sheer openness of the antagonism might make it easier for management to come to grips with difficult, cost-cutting issues such as layoffs.

New Power Technologies

Most alternative power technologies will be losers in a more competitive market, because their costs are still astronomically high compared to gas- and coal-fired generation. The high-cost alternative technologies include wind and solar power and fuel cells.

But the techniques of new power technologies will benefit from the competition necessary to lower production costs. Utilities may develop more efficient fuel cells to serve hospitals and office buildings, and build better modified jet engines to power the generation of electricity during brief peaks in demand. Perhaps NASA research on solar power cells for space stations can be spun off as more efficient and competitive alternative technology for residential electricity customers living in geographically isolated locations.

Large Companies

Mergers should favor large companies if the mergers are well-considered. Large utilities should enjoy greater economies of scale and production synergies. The larger, merged utilities, will also have greater financial resources and the ability, for instance, to spread the costs of multimillion dollar advertising programs over larger customer bases. UtiliCorp and Peco intend to do this with their EnergyOne campaign.[12] You would also expect lawyers to argue in favor of utility mergers. Morton Pierce, the head of the merger and acquisition group at Dewey Ballantine, a New York law firm that was involved in Western's acquisition of the Kansas Gas and Electric Company in 1992, thinks that merged companies such as Western and KCP&L would have lower operating costs.[13]

Consultants, however, question whether or not the current surge of merger and acquisitions activity in the U.S. utilities industry is not simply the herd instinct at work. They expect that many of the current mergers will not live up to their promise. Utilities mergers are complex and difficult to complete. To merge two large corporations requires the cooperation of often-different management and operating styles and this effort often hinders operational functions. As of February 1997, although there had been billions of dollars of announced mergers, only one large merger had been completed, that of Western and KCP&L. Critics also point out that although competitive costs are important under deregulation, they are difficult to achieve because of the large numbers of regulatory approvals the utilities must secure to do anything.

Mercer Management Consulting, a Boston firm said that only twelve mergers out of 43 completed utility mergers from 1985 to 1995 provided superior revenues and growth in operating profits.[14] This kind of result suggests that synergies in utility operations may be difficult to find.

The West: Excess Capacity

The western United States, primarily California, Oregon, and Washington, will benefit because deregulation will allow electric rates to fall while this area's huge utilities still suffer from an excess of generating capacity.

Pacific Gas & Electric: Specializing in Power Transmission

Pacific Gas and Electric Company (PG&E), the nation's largest utility, is a good example of a large western firm that is aggressively managed, but still has above-market energy rates. Although this glaring rate problem is most evident in their home state of California, their high costs do not seem impede PG&E's rapid nationwide expansion. In November 1996, PG&E purchased Teco Pipeline Company from TRT Holdings of Corpus Christi, and in December of the same year bought Energy Source Inc., a Houston-based natural gas marketer. In February 1997, PG&E offered $722 million in stock for the natural gas business of Valero Energy. These acquisitions supported PG&E's position as a major national seller of natural gas and electricity. PG&E now has "a very broad marketing reach that is literally coast to coast and up into Canada," said Robert D. Glynn Jr., PG&E's president and chief operating officer.[15]

Despite its size, PG&E has the nation's highest electric rates. In October 1996, PG&E made a move to specialize in power transmission by selling four of its electrical generating plants for $400 million to satisfy the state of California's program to open the state's utilities to competition.[16] California regulators are determined that consumers will be able to buy power from the cheapest sources beginning in 1998, and they do not care if these are out-of-state sources.[17] The state program presents one future scenario for U.S. power competition: utilities specializing in transmission might well be transmitting power for companies that they previously competed with generated in plants that they had sold to competitors. U.S. energy regulators have gone much further to deregulate power generation than they have to deregulate power transmission. There is no clear pattern for the regulation of transmission lines unless their need for regulation gives new life to the old state regulatory commissions.

PacifiCorp: Large and International

Another large west coast utility that will benefit from deregulation is PacifiCorp. PacifiCorp is an Oregon-based company that competes with PG&E. PacifiCorp is successfully building a low-cost international business in gas and cheap, coal-fired electricity generation. PacifiCorp had revenues of $4.29 billion in 1996 and a net income of $504.9 million. Its two main energy divisions, Pacific Power and Light and Utah Power and Light, provide energy to parts of Idaho, California, Montana, Oregon, Utah, Washington, and Wyoming.[18] After making two major acquisitions of power generating plants in Australia, in June 1997, PacifiCorp made a significant move toward foreign expansion with its $6 billion bid for Energy Group PLC, a large coal and electricity generating utility in Britain. If PacifiCorp is successful in working its merger with Energy Group past British regulatory resistance, it will have added superb, international, low-cost, coal and electric generating assets to its strategic portfolio.

The Northeast: Stranded Costs

The Northeast will benefit especially from new competitive prices because customers in these areas pay high rates despite excess generating capacity. With 80 percent, 70 percent, and 66 percent, respectively, of their generating produced by high-cost nuclear plants, Vermont, Connecticut and New Jersey[19] could reasonably expect to give their consumers lower electricity rates once the stranded costs of nuclear production are either paid off or refinanced. If, however, the utilities bankrupt against the stranded costs of nuclear production the rate-paying customers will benefit at the expense of utility stockholders and bondholders, and replace their local generators with out-of-state competitor utilities.

Northeast Utilities: Blocked Rates

When you discuss Northeast Utilities, you are discussing a loser, but the courts are trying to market their ratepayers into winners. When Northeast Utilities was blocked by their state public utilities commission from passing their stranded costs of generation on to their ratepayers, the customers won and Northeast's stockholders lost. This will remain true unless Northeast bankrupts against its stranded costs. If they go bankrupt, the situation is less clear.

Northeast did complain in federal court that the commission's decision would ultimately bankrupt the utility,[20] and the case may still set a precedent for how these costs may be fairly allocated. It will be interesting to see how the court system takes into account the economic cost of bankrupting utilities—or the lack of such costs. In another case, Florida Power and Light Company (FP&L) complained that they were being required to purchase electricity that was from 80 to 100 percent more expensive that what it could be by shopping around.[21] In a situation like this a court might well decide to let a utility go bankrupt, let the stockholders suffer, and allow the state's power consumers to purchase their electricity from the national grid, or from out-of-state utilities who choose to enter their territory.

Florida: The Environment

Florida Power & Light Versus Sugar Interests

Notheastern's situation is a stereotypical mess: stockholder interests must be balanced against those of ratepayers and taxpayers as courts decide who will pay how much of the stranded costs of nuclear production. The case of Florida Power & Light, however, has a more interesting factual situation. The power that Florida's contracts obligated it to buy from Okeelanta Power and Gator Generating Company was being generated by equipment designed to burn the waste products from refining sugar. This posed the additional questions of the costs of environmental pollution, waste disposal, and the synergies of two industries—the highly subsidized Florida sugar industry and the electrical utilities—as well as the basic question of whether the original investment in such waste burning equipment was a sound investment. The federal court handling the Northeast case must only decide who should bear the burden of investments in the high-cost nuclear generation equipment, although there is little doubt that the investment in nuclear generation equipment was thought to be prudent at the time it was made.

There is another interesting contrast between the Northeast case and the Florida case: Northeast is a privately owned group of utilities, whereas the bankruptcy of Okeelanta and Gator would result in the default of bonds issued by a municipality. So the courts will set precedents for whether or not the ownership of utilities with stranded costs matters. They will get to balance the equities and economics of letting corporations and bond funds go bankrupt as compared to bankrupting cities. The bonds issued by the municipalities that financed Okeelanta and Gator were held by large bond funds: Dreyfus Co., Eaton Vance Management Corp., and Franklin Resources Inc. Although the ownership of these bonds should not logically or legally be a factor in the court's decision, writers like Kingstad have graphically described the power of Wall Street in such legal and quasi-legal decisions. The power of Wall Street to approve or disallow access to future stock and bond offerings can not be overemphasized.[22]

Table 7.1 presents a selected list of corporations and their diverging financial fates based on deregulation changes and merger and acquisitions activity.

Table 7.1
Diverging Fates

Company	Bond Rating	P/E Ratio	Comment
Duke Power Charlotte, N.C.	AA-	13.3	Strong balance sheet, excellent operator of nuclear plants, growing regional economy.
Consolidated Edison of New York	A+	10.2	Conservative financing. Competitors are stymied by a shortage of power lines to send electricity into Con Ed territory.
Pacific Gas & Electric San Francisco	A	11.7	Its rates are high. But California regulators will allow it to recover the costs of building expensive nuclear plants.
Texas Utilities, Dallas	BBB+	12.5	Healthy regional economy. Has older nuclear plants that were less costly to build.
Niagara Mohawk Power, Syracuse	BB-	6.9	To avoid a Chapter 11 filing, attempting to escape obligations to buy power at far above the market price.

Source: Standard & Poor, *Business Week*. Ratings and PE ratios taken in December 1996.

LOSERS

Nuclear Power Plants

Nuclear power plants are clear losers, because they were costly to build, and they're not cheap to operate. But nuclear power plants, despite the engineers who love them, are only assets, and most of the utilities that own them will survive by replacing them with more efficient gas-fired plants. The future of these plants, then, at least in the United States, is to be mothballed, abandoned, and contracted out to the few groups of experts who seem to be able to run them efficiently. But even if nuclear generating plants are properly maintained, safe, and run efficiently by good management, the plants will lose the efficiency race to coal and gas.

Investors

Widows and orphans and other conservative investors need to watch out for the falling bond ratings of utilities that own nuclear plants, as the Northeast Utilities case makes plain.[23] Stockholder dividends also are at risk if regulatory commissions choose the wrong method for refinancing stranded costs in nuclear reactors, because refinancing at rates too high or at maturities too short will leave the utilities vulnerable to lower cost competition. Even healthy utilities have cut dividends to invest in growth assets leading to better competitiveness than nuclear generation plants will allow.

The worst case scenario for utility bondholders is a utility filing for bankruptcy. The bankruptcy filing of the huge Florida Power & Light roiled financial markets and threatened default on $288 million of high-risk municipal bonds that had been used to finance the generation plants of Okeelanta and Gator. If the courts allowed FP&L bankruptcy protection, that action would probably result in a default on the bonds and either a full or partial loss on the part of anyone owning the bonds. And the owners of the bonds include the shareholders of the large funds that invested in the bonds, firms like Dreyfus Company, Eaton Vance Management, and Franklin Resources.[24]

Homeowners

Homeowners will benefit from the deregulation of the supply of electricity because they will be able to buy electricity from the lowest cost sources through their local electrical utilities. If a local utility is not able to produce electricity as cheaply as national power-marketing electricity generators, in the future the local utilities will be able to contract out the supply of cheap electricity and make their profit on its transmission.

Although homeowners and their utilities can shop for a supply of electricity, that is not true of the transmission wires that bring them the electricity from competing power plants. These wires will remain a regulated monopoly for years to come, as mergers between competing power retailers produce one or two transmission companies per region.[25]

Some localities, such as Lynbrook, Long Island, may choose to form their own municipal electrical utilities rather than pay higher prices of a utility strapped with the costs of abandoned nuclear reactors.[26] The FERC regulations allow the startup municipal utilities to condemn the lines of the previous power company in order to deliver the municipal utility's electricity.[27]

Industrial Customers are
Power Consumers

Homeowners will benefit less than large industrial customers of utilities. Even as the state of New York announced the breakup of their Edison utility, the utility announced that it would give deep rate cuts to large industrial customers but not to residential customers.[28] And on Long Island, as a part of the state plan to restructure the Long Island Lighting Company (LILCO), New York's governor George Pataki's administration planned on lowering manufacturing electricity rates by 25 percent over five years but admitted that small business and residential consumers would receive much less.[29]

Industrial customers have already begun to reap the rewards of competition by purchasing electricity in large quantities and through long-term contracts. Although these much lower prices attract industry and make business more competitive, these customers say they are having difficulty in comparing the complex contract provisions and assessing the dependability of the service that they are committing to for such long periods of time.

The utilities, on the other hand, are becoming expert at crafting contracts, and would like to reduce their risky exposure to changes in energy prices by matching up their long-term commitments for natural-gas[30] to fuel their gas-fired generating plants with long-term contracts with their industrial customers. Texas-based Enron, for example, has invested $250 million in a computer program that will develop the information that is the raw material for drafting these contracts, and promoted its energy trading expert, to the position of their number two executive, showing how important they think energy trading skills are.[31]

Reliability of Electrical Supplies

The reliability of electrical supplies may be a casualty, caution utilities executives, because with generators competing to transmit electricity over the same wires, the chance of accidental overloads rises. It may also be necessary to put wires in a region under an "independent system operator," because you can't store electricity and a regulatory mechanism is necessary to prevent system overloads from electricity suppliers.[32]

An optimistic, objective, market-oriented observer of the coming deregulation might suspect that these objections and fears are merely the carping of utilities yearning for the status quo. To quote Milton Friedman, "[N]obody likes competition," or as Harry Levitt, president of the Lynbrook Chamber of Commerce said, "I'd love to see what FERC says [about this]."[33] Wall Street analysts feel optimistic about future reliability, and, at least in the gas segment of the industry, "[T]he deregulation of the gas markets lowered prices and, together with the environmental benefits of gas, increased the scope, quality and reliability of services and stimulated demand.[34] Wall Street sounds optimistic.

Taxpayers

Taxpayers may get stuck with the cost of refinancing stranded utility investment because the government is likely to write down billions of dollars in loans to strapped rural electric cooperatives. California is now the bellwether state in negotiations that may allow utility customers to pay for all of the state's utilities' stranded costs. To sweeten the deal, a 10 percent rate reduction has been put up front, with 10 percent more in the year 2000. But the battle in California hasn't been settled yet and customers may not agree with this plan.

Holders of Securities

Utility securities offer much more risk and reward under deregulation than they did when they were regulated monopolies. State power commissions will gradually refuse to guarantee stockholders' dividends and bondholders coupon payments unless they fund investments in transmission lines. This is already beginning to result in bankruptcy situations for utilities with power plants that built in high, uncompetitive costs. These situations threaten potentially disastrous decreases in shareholder and bondholder value.

The other side of the equation is that the winners of the merger-competition game unleashed by the decontrol of energy supplies will be big winners, along with their stockholders. Will bondholders benefit too? Maybe, but certainly not as much as the stockholders will. Bondholders, because of the nature of their security, find value where certainty is assured. Stock, on the other hand, tends to increase in value, relative to risk, when cash flows become more volatile so the stocks of synergistic utilities will show handsome capital gains. Options pricing models virtually guarantee these results. Under deregulation cash flows will be more volatile. On the up side, strong utilities will see cash flows fluctuate more because long-term energy contract values will fluctuate with market energy rates. On the down side, weak utilities are more likely to go bankrupt. So for both strong and weak utility investments, there will be more risk, and the volatility of the new market will treat equity investments better than bond values, and synergistic mergers better than bootstrapped mergers.

The investors who are hurt most by this situation are older investors, smaller investors, or investors who, for some reason, are not able to bear increased risk. But return will still be proportional to risk in utilities investments, although equity investments can expect to do better than investments in debt.

Freebooters

The new, deregulated markets for energy will create a new category of entrepreneur—freebooters. Freebooters are wholesale marketers of energy who own neither wires nor power plants. These entrepreneurs may be allowed to use their competitor's wires much as the Bell System was forced to let MCI use their local phone lines. These marketers are expected to provide customized pricing plans that package electricity with other products—say telephones.[35]

ENERGY-TRADING INTERNATIONAL POWER MARKETERS

Pacific Gas & Electric Company

The nation's largest power company, Pacific Gas & Electric Company (PG&E), owns the plants that produce the commodity they trade. But like freebooters, power-marketing utilities like PG&E actively trade energy contracts. Forced to divest itself of four of its California generating plants, some of the highest-cost generating companies in the United States, PacifiCorp has moved aggressively across the United States by acquiring the Teco Pipeline Company, Energy Source Inc., and Valero Energy Corporation, all natural gas marketing facilities. These acquisitions are shown in Table 7.2.

Table 7.2

Pacific Gas & Electric Company Expands Nationally

Date	Acquisitor	Target Company	Cost
Nov 1996	Pacific Gas & Electric Co.	Teco Pipeline Co. from TRT Holdings (Corpus Christi)	?
Dec 1996	Pacific Gas & Electric Co.	Energy Source Inc. (Houston)	?
Feb 1997	Pacific Gas & Electric Co.	Valero Energy Corp.	$0.723 B

Sources: Benjamin A. Holden, "PG&E Agrees to Buy Unit from Valero," *Wall Street Journal*, February 3, 1997, pp. A1, A3; and Agis Salpukis, "Enron Utility Merger Approved in Move to Expand Nationwide," *New York Times*, February 27, 1997, pp. D1, D4.

PacifiCorp

Like PG&E, PacifiCorp is an energy-trading international power marketer. But unlike PG&E, PacifiCorp, expanded internationally. A large coal and gas-fired generator, PacifiCorp began acquiring $3.46 billion of privatized, deregulated Australian power assets in 1997. It then punctuated this rapid expansion by making a $ 6 billion offer for Britain's Energy Group. PacifiCorp seemed determined to acquire Energy's coal production and transform itself into an global en-

Table 7.3

PacifiCorp Expands Internationally

Date	Acquisitor	Target Company	Cost
Nov 1995	PacifiCorp (Portland) electricity telephones	Powercor (Australian) electricity	$1.6B
Aug 1996	National Power PLC (U.K.), Commonwealth Bank Group Australia), PacifiCorp (U.S.) and Destec Energy Inc. (U.S.)	Hazelwood Power Station (Australian)	$1.86 B
Jun 1997	PacifiCorp	Energy Group, PLC	$6B

Sources: Benjamin A. Holden, "PacifiCorp to Buy Utility in Australia," *Wall Street Journal*, November 16, 1995, p. A3; Benjamin A. Holden, "PacifiCorp and Destec Join Group to Pay $1.86 Billion for a Utility in Australia," *Wall Street Journal*, August 5, 1996, p. A4; James P. Miller and Steven Lipin, "CalEnergy Launches Another Hostile Bid," *Wall Street Journal*, July 16, 1997, pp. A1, A3, A4; and Agis Salpukas, "$1.9 Billion Hostile Bid For Utility," *New York Times*, July 16, 1997, pp. D1, D18.

ergy utility specializing in coal production, power generation, and energy trading and sales. These acquisitions are shown in Table 7.3. Since the utilities they acquired are private, deregulated, and produce a higher rate of return than Pacifi-Corp's domestic assets, it is difficult not to call PacifiCorp a winner.

Purchase Power Contracts

The state of New York requires its utilities to buy the output of small, high cost independent electricity generating firms under a 1978 federal law passed after the OPEC oil cartel roiled world oil prices in 1975. This law was an attempt to stimulate domestic energy production and staunch U.S. dependency on OPEC oil, but in retrospect, the law can be seen to have been ill-conceived. The purchase prices mandated by the law were extrapolated from the high oil prices experienced during the initial, scary stages of the introduction of OPEC. These electricity prices are now as much as three times higher than current market prices, forcing Niagara Mohawk to pay from 6 to 10 cents per kilowatt hour when they could buy electricity in the market for 2 to 3 cents per kilowatt hour. These price premiums have caused Niagara Mohawk annual overpayments of as much as $600 million.[36]

Nationwide, purchase power contracts cover just 7 percent of generated electricity, but they account for $65 billion in stranded costs, nearly a quarter of the utility industry's total stranded costs.[37] But estimates of the value of stranded costs varies a lot. One report finds that nationwide, total stranded costs amount from $45 to $400 billion, with almost 40 percent accounted for by nuclear costs.[38] A breakdown of the sources of stranded costs is shown in Table 6.2.

Buyouts to Achieve Competitiveness

In March 1997, Niagara Mohawk announced that it had agreed to a $4 billion buyout of the long-term contracts of most of its independent producers.[39] Mohawk estimated that the buyouts would save the utility $5 billion over the next fifteen years, and end or restructure forty-four of their long-term power contracts with nineteen independent power producers.[40] The buyout of the contracts would be a part of New York's preparation for deregulated, competitive power markets by 1998.

Although most of the buyout were to be financed with bonds, $400 million were paid in cash. The remaining $3.2 billion would be financed with debt. Niagara Mohawk also agreed to issue forty-six million shares of stock to the independent producers, giving them a future stake in the utility.[41] The deal had the effect of allowing the independents to exchange their contracts for cash and a share in Niagara Mohawk. Dan Scotto, a utility analyst for Bear, Stearns & Company, said that the agreement provided a windfall for the independent power producers, because most of their contracts would be paid off, but they would still own the plants and be able to sell power under new contracts, although at lower rates, with Niagara Mohawk.[42]

The debt portion of the buyout was similar to the way that LILCO refinanced their stranded costs of nuclear generation. The long-term bonds funding the buyout would be repaid in smaller amounts over a longer period of time than the contracts would have been. This means that Niagara Mohawk's current costs will be lower, allowing their prices to be lower and more competitive.[43] The shares in Niagara Mohawk let the independents identify financially with the large utility, but the shares also dilute the interests of the current stockholders and must, therefore, be approved by the stockholders. Niagara Mohawk's stockholders are being asked to accept an adverse short-term situation in return for future profitability.

Mohawk is the first utility to attempt to buy out all of its independent producer contracts at one time. The deal is much the same as if Mohawk had merged with their many, small, non-competitive power producers. Still, the large utility hopes to be competitive and to pay off the $3.2 billion of debt by selling power at market rates. Although the buyouts pushed up state electricity rates, the buyout deal may provide a model for Con Edison to buy out their independent contracts.

Mohawk is an upstate New York utility serving 1.5 million customers in the western lobe of the state around Buffalo. This leaves New York City and the central and northwest-southeast diagonal of the state for Con Edison. Niagara Mohawk felt that they had to arrange the buyouts now in order to avoid a case of slow Chapter 11 bankruptcy under deregulated competition.[44]

Good-bye Subsidies

Deregulated competition will probably wipe out a number of programs that have the flavor of developmental subsidies:[45]

- Con Edison's New York purchase power agreements: agreements to buy power from independent power generators even if it is more expensive.
- Con Edison's programs to subsidize electricity for low-income customers—stranded costs in a competitive market.
- "Alternative" energy programs: purchase agreements to support solar energy for example.

The Nuclear Power Puzzle

The gains and losses produced by the $86 billion that utilities have stranded in nuclear generation costs will touch everybody—ratepayers, taxpayers, stockholders, bondholders, managers, and Wall Street. These costs represent 43 percent of the total stranded costs of utilities, surpassing the 25 percent of costs stranded in purchase power contracts.[46] The breakdown of these costs is shown in Table 6.2.

More than 100 nuclear plants generate nearly 25 percent of the nation's electricity, so there is heavy reliance on nuclear power in some areas, and every region of the country shares the problem of stranded nuclear costs.[47] These stranded costs fall particularly heavily on Vermont, Connecticut, Illinois, New Jersey, and South Carolina, because all of these states produce at least 66 percent of their power with nuclear plants.

Table 7.4
Winners and Losers Under Various Plans
to Refinance Stranded Costs

Strategy/Event	Result	Because

Playing for Time

Strategy/Event	Result	Because
1. Refinance with bonds	New bonds are riskier, ratepayers rates are reduced some.	Ratepayers pay stranded costs over a longer period of time with a gradual transition away from dependence on nuclear utilities: Connecticut, New Jersey, South Carolina, Vermont, Illinois, Chicago.
2. Refinance with tax exempt bonds	Some taxpayers lose, rates are reduced some.	Increases the tax burden on some taxpayers not relieved of costs.
3. Utility in Chapter 11 bankruptcy	Stockholders, bondholders and ratepayers share the cost	Stock falls in value, bond coupons are renegotiated, and rates go up some.

Replacing the Problem

Strategy/Event	Result	Because
4. Utility goes bankrupt, Chapter 10	Stockholders and bondholders lose. Ratepayers win.	Valueless stock and defaulted bonds. New low cost firms charge lower rates.

Managing the Problem

Strategy/Event	Result	Because
5. Utility is merged with another, stronger utility	The stockholders and bondholders of both companies suffer as do ratepayers	Stranded costs are averaged over both firms causing fall in stock and bond values. Rates go up some.
6. Utility is merged with a nuclear specialist	Everybody wins.	Better management brings down nuclear costs to the lower firm average—lower rates.
7. Utility is bought by a nuclear specialist	Everybody wins.	Better management brings down nuclear costs to the lower firm average—lower rates.

But some of these regions, like the Chicago area, manage their nuclear generation better than other areas. Chicago's Commonwealth Edison supplies 70 percent of the area's electricity from twelve well-managed nuclear plants.[48] Edison has some nuclear plants that are efficient enough to sell on the market, although other of their plants are among the most inefficient, and some plants face huge renovation expenses. Edison has accumulated $10 billion of nuclear-related debt, and it welcomes competition, but it feels that it deserves time and the financial breaks from regulators that will allow their preparation for it.[49]

Table 7.4 shows some of the strategies used to deal with stranded nuclear costs.

PLAYING FOR TIME

Table 7.4 suggests three basic cures for stranded costs: playing for time, replacing the problem utility, and managing the problem nuclear plants. Playing for time has been the most popular solution in New York. New York has encouraged the refinancing of the $6.5 billion of stranded nuclear costs of LILCO and the expected $14 to $25 billion costs of refinancing it, as well as approving the $4 billion buyout of Niagara Mohawk's purchase power contracts. The state still needs to approve the billions of dollars of financing necessary to refinance Con Edison's old debt and buy out its power purchase contracts. The state hopes that the refinancings that spread out the payment of these stranded costs will reduce the current costs of the utilities enough to let their prices be competitive.

Replacing the Problem

When a state allows a utility to go bankrupt, this is generally equivalent to inviting the entry of a larger, more efficient national firm to take over the bankrupt utility's abandoned territory. To avoid such a bankruptcy situation, the New England Electric Company recently put its eighteen fossil-fuel fired generating plants up for sale. CalEnergy, Southern, and Duke Power Company were interested in buying the plants, but Pacific Gas & Electric outbid the other companies, paying $1.59 billion, a 45 percent premium over the plants' book value of $1.1 billion.[50]

Hartford-based Northeast Utilities shut down its troubled Connecticut Yankee plant fourteen years ahead of schedule. But when it closed its inefficient plants it had to lay off large numbers of employees, and burdened New Hampshire with huge tax losses.[51] If the federal court disallows Northeast Utilities' 1997 rate increases, in the case now pending, this group of utilities may be replaced and find their New England territory taken over by an out-of-state power-marketing firm.[52]

Whereas the New Hampshire public utilities commission put Northeast Utilities under enormous pressure, California and Pennsylvania are giving their utilities a number of years to adjust to deregulation, thus softening the transition to competition. They are doing this mostly with freezes or reductions in overall electric rates. But California and Pennsylvania do not have the serious problems with stranded nuclear costs that Northern Utilities does.[53]

Out-of-state replacement is not the inevitable result of bankruptcy, however. If Florida allows the high-cost Okeelanta and Gator generating companies to go bankrupt, their territory would presumably be taken over by in-state Florida Power and Light Company (FP&L), which has claimed that honoring its mandated contracts with Okeelanta and Gator would bankrupt it. Florida courts will probably decide in favor of FP&L because it is the low-cost producer, or they will play for time by allowing FP&L to renegotiate their contracts with Okeelanta and Gator. The renegotiated contracts would then put pressure on Okeelanta and Gator to prove that their waste-burning generation technology is efficient—or they would go bankrupt and FP&L would buy their power on the grid at market rates.

The replacement-through-bankruptcy solution is appealing in cases of poorly managed utilities because, although it punishes the utility's investors, the result seems just, the utilities industry gets an invigorating whiff of Schumpeterian competition, and there is an immediate reduction in electric rates.

Managing the Problem

Managing the problem is appealing because everybody wins, even the specialist managers. Presumably the specialist managers are very fond of nuclear production technology, because few have learned to manage it so well. And that is the problem: good nuclear generation managers are *scarce*. This must be true or good nuclear management would have already emerged as the happy, universal solution—and there would be no stranded nuclear production costs.

Despite the fact that better management has been shown to significantly reduce the costs of nuclear generation, utilities with heavy investments in nuclear reactors are still going to be riskier than those using coal- and gas-fired generation. This is because the new coal- and gas-fired generation technologies have the double-barreled advantage of lower cost technology and lower cost fuel.

New York's Governor George Pataki's administration pioneered a variation on the themes suggested in Table 7.4 when it stratified the rates proposed in a plan to restructure Con Edison. Con Edison's rates are approximately 30 percent above market rates. So to protect and stimulate the local economy, the New York governor chose to leave residential rates virtually unchanged and cut electricity rates to businesses by as much as 25 percent. This kept the costs high for residential consumers, the market segment which was least able to negotiate their rates down.

The New York plan is a trickle-down, business-first approach to consumer welfare. The state argues that stimulating business stimulates the state economy and indirectly improves the welfare of everybody: businesses, new employees, and, eventually, ratepayers. As a medium-term strategy this is very appealing and will probably be a popular solution nationwide.

The trickle-down approach suggests that certain localities, like Lynbrook, Long Island, have done exactly the right thing when they opt out of the services of very expensive utilities, like LILCO. Pataki's plan may stimulate the state's economy, but it allows high LILCO electric rates to depress Lynbrook. In this situation, FERC encourages local municipalities like Lynbrook to opt out of state plans that hurt them. FERC encourages municipalities to start up their own small, local, competitive, entrepreneurial utilities. FERC wants these municipal utilities to compete with various large utilities that have notorious records of inefficiency and high rates. FERC's rules make this kind of municipal competition possible by allowing the municipalities to condemn their installed power lines in order to deliver their new utility's low-priced electricity. Being able to start up a municipal utility gives municipalities a major bargaining tool against state plans that may be unjust.

NOTES

1. Peter Coy and Gary McWilliams, "Electricity: The Power Shift Ahead," *Business Week*, December 2, 1996, pp. 78-82.

2. Agis Salpukas, "New Choices for Natural Gas. Retailers Find Users Puzzled as Industry Deregulates," *New York Times*, October 23, 1996, pp. D1, D4.

3. Stephen W. Bergstrom, senior vice president at NGC Corp., a Houston-based natural gas and electricity marketer, in Coy and McWilliams, "Electricity: The Power Shift Ahead," pp. 78-82.

4. Jon Erik Kingstad, "Merger Menace: Holding Companies and Overcapitalization," *Public Utilities Fortnightly*, January 15, 1996, pp. 42-45.

5. Dana Canedy, "Suitor Drops $1.9 Billion Bid for Utility," *New York Times*, August 16, 1997, pp. 35-36.

6. Canedy, "Suitor Drops $1.9 Billion," pp. 35-36.

7. *Ibid.*

8. Richard Roll, "The Hubris Hypothesis of Corporate Takeovers." *Journal of Business*, Vol. 59 (April 1986), pp. 197-216.

9. Agis Salpukas, "Now Comes Hard Part in Utility Deal," *New York Times*, February 10, 1997, p. D2.

10. *Ibid.*

11. *Ibid.*

12. Benjamin A. Holden, "UtiliCorp and Peco, Aided by AT&T, to Launch One-Stop Utility Service," *Wall Street Journal*, July 24, 1997, pp. A1, A3; and "Home-Services Alliance Forms," *New York Times*, June 25, 1997, pp. D1, D5.

13. Salpukas, "Now Comes Hard Part," p. D2.

14. *Ibid.*

15. Benjamin A. Holden, "PG&E Agrees to Buy Unit from Valero, "Wall Street Journal*, February 3, 1997, pp. A1, A3.

16. Agis Salpukas, "Pacific Gas and Electric to Sell 4 Power Plants," *New York Times*, October 23, 1996, p. D4.

17. *Ibid.*

18. Agis Salpukas, "Says Pacificorp Considers a Bid," *New York Times*, June 11, 1997, pp. D1, D2.

19. Barnaby J. Feder, "The Nuclear Power Puzzle; Who Will Pay for a Generation of Expensive Plants?" *New York Times*, January 3, 1997, pp. D1, D3.

20. Agis Salpukas, "Northeast Utilities Sues to Block Move by New Hampshire," *New York Times*, March 4, 1997, p. D8.

21. Charles Gasparino, "Electric Giants File for Bankruptcy Protection," *Wall Street Journal*, May 19, 1997, pp. A1, A3.

22. Kingstad, "Merger Menace," pp. 42-45.

23. Salpukas, "Northeast Utilities Sues," p. D8.

24. Gasparino, "Electric Giants File," pp. A1, A3.

25. Coy and McWilliams, "Electricity: The Power Shift Ahead," pp. 78-82.

26. Andrea S. Halbfinger, "Study in Contrast, Lynbrook Village Meeting has Music Cute Kids and Conflict," Lynbrook Local News, March 6, 1997, pp. 1, 3, 26.

27. Jo-Anne Taormina, "Letter from Lynbrook Village Community Relations: Vote on a Proposal to Allow the Village to Condemn a Portion of Lilco's Transmission Wires," unpublished letter.

28. Richard Perez-Pena, "Pact Reached to Break Up Con Edison," New York Times, March 14, 1997, pp. B1, B4.

29. Agis Salpukas, "When Electricity Goes Private, Deregulation May Change New York Power Authority," New York Times, July 11, 1997, pp. D1, D2, D3.

30. Daniel Pearl and Peter Fritsch, "Deep Pockets: Natural Gas Generates Enthusiasm and Worry In Oil-Soaked Mideast," Wall Street Journal, August 11, 1997, pp. A1, A8.

31. Agis Salpukas, "Enron Names Trading Chief as President of Company; Appointment Reflects the Changing Market," New York Times, December 11, 1996, pp. D1, D6.

32. Coy and McWilliams, "Electricity: The Power Shift Ahead," 78-82.

33. Halbfinger, "Study in Contrast," pp. 1, 3, 26.

34. Judith B.Sack and Robert L. Chewning, "Global Electricity Strategy: My Two Cents Worth (or The Sustainable Price of Power)," Morgan Stanley, International Investment Research, February 6, 1997.

35. Coy and McWilliams, "Electricity: The Power Shift Ahead," pp. 78-82.

36. Gordon Fairclough, "Niagara Mohawk to Pay 19 Producers $4 Billion to Alter, End Energy Pacts," Wall Street Journal, March 11, 1997, pp. A1, A3; and Agis Salpukas, "Utility Seeks to End Costly Pacts With Power Suppliers; A Move to Shed Contracts That Have Raised Power Rates," New York Times, March 11, 1997, pp. D1, B8.

37. Jeff Bailey, "Niagara Mohawk Plan is a Small Step Toward Easing Utilities' Power Woes," Wall Street Journal, March 12, 1997, pp. A1, A4.

38. Feder, "The Nuclear Power Puzzle," pp. D1, D3.

39. Fairclough, "Niagara Mohawk to Pay," pp. A1, A3; and Salpukas, "Utility Seeks to End Costly Pacts," pp. D1, B8.

40. Ibid.

41. Ibid.

42. Agis Salpukas, "Niagara Deal With Independents Could Reduce Price of Electricity," New York Times, July 11, 1997, pp. B5, D1.

43. Ibid.

44. Ibid.

45. Feder, "The Nuclear Power Puzzle," pp. D1, D3.

46. Bailey, "Niagara Mohawk Plan," pp. A1, A4.

47. Feder, "The Nuclear Power Puzzle," pp. D1, D3.

48. Ibid.

49. Ibid.

50. Charles V. Bagli, "PG&E Will Buy 18 Power Plants in New England," New York Times, August 7, 1997, pp. D1, D2.

51. Feder, "The Nuclear Power Puzzle," pp. D1, D3.

52. Salpukas, "Northeast Utilities Sues," p. D8.

53. Feder, "The Nuclear Power Puzzle," pp. D1, D3.

Chapter 8

The Global Spread of U.S. Utilities

UTILITIES SWAP VERTICAL FOR HORIZONTAL MONOPOLIES

The U.S. utilities industry is rapidly dismantling its vertically integrated power monopolies and replacing them with horizontal monopolies based on new specializations and the larger markets that these specializations make possible.[1] Decisions in state courts, as well as by state and federal regulatory agencies are making these changes by chipping away the guaranteed prices and low costs of capital upon which utilities have relied. The changes in regulations are being inspired by the Federal Energy Regulatory Commission (FERC).

By deregulating the market for the supply of electricity and streamlining their procedure for approving mergers, FERC has been leading American utilities in their preparations for operating in deregulated, competitive markets. State and federal laws supporting regulated utility monopolies are being reinterpreted or repealed. Without state commissions to regulate and protect their monopolies, the utilities have few incentives to keep the production and transmission of electricity under one roof. Without guaranteed monopoly prices and low-risk costs of capital, utilities must cut costs and increase profits by developing new power specializations: gas-fired generation, coal-fired generation, gas-to-electric convergence and energy arbitrage, power transmission and energy marketing, or even one-stop power marketing. The current crop of mergers indicates that such specializations are already a key component of new utility strategies.

Specialization and Market Dominance

Financial theory approves of the tendency toward corporate specialization because the theory dictates that corporations cannot do better for their stockholders than their stockholders can do for themselves. The utilities industry, like other industries, has found that stockholders can diversify their stock portfolios much more easily than the utilities can diversify their portfolios of investment assets. So utilities have begun to specialize in what they do best, which is what they can do most profitably which is to sell gas and electricity. The deconglomerization that this specialization encourages has caused utilities to spin off businesses that were not related to their core competencies in energy production.

At the same time that they began to specialize, utilities began to merge with other utilities to extend their markets. The extension of specialties through mergers has left the utilities industry with fewer, larger corporations. In most cases this has cut operational costs, decreased utility rates, increased profits, enhanced shareholder wealth, and increased competition. Curiously, the drive for competitive efficiency has also led utilities to merge more often with other non-contiguous utilities—and regulators have approved of this trend because it seems to encourage competition.

Specialization is not unique in the utilities industry; it was a major theme in the recent phenomenon of corporate downsizing. Diversified corporate conglomerates have almost disappeared in the United States. The downsizings were also combined, in almost every case, with a search for core competencies and larger markets. Although downsizing resulted in trimmer, more profitable businesses, even when they began their downsizing many of these corporations were multinational giants; now that these trimmer multinationals have completed their downsizing they find that their increased market reach has made them truly global businesses.

Part of this new global reach is the result of utilities striving to master their new competitive environment by increasing their market share—in other words, a strategy of market domination. Corporate strategies of competition and market dominance result, quite naturally, in horizontal monopoly or oligopoly. But oligopoly and monopoly are only goals that may be reached in the face of industry competition and regulatory oversight.

Mergers Specializing in Coal- and Gas-Fired Generation and Transmission

As vertically integrated utilities transform themselves into market-grabbing oligopolists in search of horizontal market monopolies, more and more of these firms find themselves specializing in the technology of gas-fired electricity generation. Gas-fired generation is the low-cost generating technology of the coming era of free-market utilities. But specialization in this new technology is complicated by the concept of energy convergence. Gas can be converted into electricity or gas can be converted into oil, and oil can be converted into electricity. This means that electricity contracts can be traded for contract energy equivalents of gas and oil. Part of the process of specialization in the appropriate technology involves accumulating a knowledge of emerging energy markets, energy commodities, energy finance, and the development of a global enterprise that can apply this knowledge.

Utilities are investing in these technological specializations and competencies in order to lower the energy generation costs of their somewhat phlegmatic corporations. The utilities feel keenly that they must prepare for more competition in deregulated energy supply markets, and their high utility rates prove to everyone that they are not yet ready. In fact, deregulation is taking place primarily in order to lower the power rates.

Under FERC's new regime, only the transmission end of the business—the sales end—will remain regulated. Because of the enormous investment in power lines, this end of the industry will change more slowly, if ever, to a competitive market, and it is tempting to think that the utilities that choose to specialize in the continuing regulated monopolies of power transmission are the less interesting firms.

The deregulation of the markets for power supply are causing electricity, gas, and other power products to be treated as commodities. This means that price competition for electricity contracts will be based on low-cost generation.[2] Companies with low generation costs will win more sales contracts than those based on aging, high-cost nuclear generating plants. Horizontal oligopolies, or something approaching them, are expected to be better competitors in this environment for a number of reasons, and are being warily tolerated by both state and federal utility regulators, as well as by U.S. antitrust law.

Specific Advantages of Horizontal Utility Monopolies

Horizontal monopolies have larger numbers of customers, and these larger numbers—by simply having a huge market—produce economies of scale. These economies of scale confer advantages on power-marketing utilities in a variety of ways (companies that exemplify these advantages are noted in parenthesis):

- Gas-to-electricity mergers unite the expertise to cross-market gas and electricity—especially to large commercial customers (Enron[3]).
- Gas-to-electricity mergers create a market niche for highly efficient electricity producer (The Duke Power Company[4]).
- Monopolies on unusual technologies can create market niches such as that for geothermal electricity production (CalEnergy[5]).
- The pairing of other products sold to residential energy customers in a product line integrated with nationally marketed, brand name electricity can create one-stop, power marketing firms in either gas or electricity (PacifiCorp[6] and Enron[7]).
- Dealing nationally in wholesale energy markets while owning no sources of energy or having any specific geographical customer base brings together the expertise and capital necessary to become a "freebooter."[8]
- Larger companies spread the costs of refinancing their stranded costs of nuclear power

plants over a much larger customer base in order to reduce costs and stay competitive.

- Mergers produce scale economies of management (Entergy becomes the contract manager for Maine nuclear plants[9]).
- Larger companies spread the costs of operations and engineering staffs across a larger number of customers. This helps utilities improve their safety records and develop knowledge that can be sold through the managerial consulting contacts of their subsidiaries.[10]

MERGERS, MONOPOLIES AND MARKET DOMINATION

The United States

Although there is a great deal of controversy over whether or not the billions of dollars of utilities mergers in the United States from 1992-1997 will create more operational efficiency than they cost in terms of market domination,[11] U.S. utilities have certainly been merging at a record rate. Table 8.1 shows almost $46 billion of mergers in U.S. utilities with other U.S. utilities during the period from 1992 to 1997.

Pacific Gas & Electric Company

The largest utility in the United States, the California-based Pacific Gas & Electric Company (PG&E) has taken a conservative approach to market specialization by investing its future in the regulated area of power transmission.

By allowing PG&E to specialize in the transmission and marketing of electricity, California regulators are allowing them to pay off their stranded costs in nuclear electricity generation by continuing to charge above market rates to their California electricity customers in a state with some of the highest electricity rates in the United States. In its home base market, PG&E has opted retain its partial regional monopoly in the regulated power transmission business while it retreats from the competitive power generation market where it has higher costs and is less competitive.

But PG&E is specializing in power transmission only as a California state strategy. Although it has chosen the safe, regulated, home-state transmission markets, PG&E's nationwide strategy appears to be to acquire low-cost gas and electricity generation capacity and to sell this gas and electricity at competitively low prices nationally. To get out-of-state business, PG&E has acquired three out-of-state gas companies in 1996 and 1997. Table 8.2 shows these acquisitions. Outside of California, PG&E is viewed as a power marketer of gas and electricity. The fact that PG&E's national strategy is more aggressive than their home-state strategy suggest that they may well return to the competitive electricity generation market in California after paying off their stranded costs.

Table 8.1
U.S. with U.S. Utilities Mergers

Date	Acquisitor	Target	Cost ($billions)
1992	Kansas Gas & Electric (now Western Resources Inc.)	Kansas Power & Light (Kansas)	$1.700
Oct 1995	Baltimore Gas and Electric	Potomac Electric Power Co.	$5.000
Nov 1995	Centralsouth West Corp. (Dallas)	Seaboard PLC	$2.530
Nov 1995	Entergy (New Orleans)	CitiPower Ltd.	n.a.
Nov 1995	Texas Utilities Co. (Dallas)	Eastern Energy	$1.550
Nov 1995	Houston Industries (Houston)	NorAm Energy Corporation	$2.400
Nov 1995	Allegheny Power System Inc.(NYC)	Duquesne Light Co. (Pittsburgh).	$0.170
Jan 1996	UtiliCorp (Kansas City)	Kansas City P&L (Kansas City)	$1.350
Feb 1996	American Water Works (Pa)	Pennsylvania Gas and Water Co.'s water utility	$0.262
Mar 1996	Western Resources Inc. (Kansas City)	Kansas City Power and Light Co. (Kansas City)	$1.700
Jun 1996	Enron Corp. (Houston)	Cajun Electric Power Coop. (New Orleans)	$1.040
Jun 1996	Southern Co. (Atlanta)	Cajun Electric Power Coop. (New Orleans.)	$1.000
Jun 1996	NRG and Ziegler Coal Holding (Chicago)	Cajun Electric Power Coop. (New Orleans.)	$1.070
Jun 1996	Southestern Electric Power (Shreveport)	Cajun Electric Power Coop. (New Orleans.)	$0.900
Jun 1996	NRG and Ziegler Coal Holding (Chicago)	Cajun Electric Power Coop. (New Orleans.)	$1.070
Jul 1996	Noble Affiliates Inc. (Tulsa)	Public Service Enterprise Group (NJ)	$0.775
Jul 1996	Enron Corp. (Houston)	Portland General Corp. (Portland)	$2.100
Aug 1996	Interstate Energy Corp. (Dubuque)	WPL Holdings (Madison), Interstate Power Co. (Dubuque), IES Industries (Cedar Rapids)	$1.900
Aug 1996	Houston Industries (Houston)	NorAm Energy Corporation	$2.400

Table 8.1 (continued)

Aug 1996	Delmarva Power and Light Company (Baltimore)	Atlantic Energy Inc.	$0.968
Aug 1996	Cinergy (Cincinnati)	Williams Cos. (Tulsa) and PanEnergy Corp. (Tulsa)	n.a.
Aug 1996	MidAmerican Energy Co. (Des Moines)	IES Industries (Cedar Rapids)	$1.200
Aug 1996	Minnesota Power & Light Co. (Minneapolis)	Adesa Corp.	$0.220
Aug 1996	United Cities Gas Co. (Tenn.)	Atmos Energy Corp. (Dallas)	n.a
Sep 1996	Ohio Edison Co. (Akron)[25]	Centerior Energy Corp. (Independence)	$1.610
Mar 1996	Western Resources Inc. (Kansas City)[26]	Laidlaw Inc. (Toronto)	n.a.
Nov 1996	Duke Power Company (Durham)	PanEnergy (Houston)	$7.700
Nov 1996	Pacific Gas & Electric Co.	Teco Pipeline Co. from TRT Holdings (Corpus Christi)	n.a.
Dec 1996	Pacific Gas & Electric Co.	Energy Source Inc. (Houston)	n.a.
Feb 1997	Pacific Gas & Electric Co.	Valero Energy Corp.	$0.723
Feb 1997	NGC Corporation	Destec Energy	$1.270
Feb 1997	AES Corporation (Arlington)	Destec Energy	$1.270
Jul 1997	CalEnergy (Omaha)	New York State Electric & Gas	$1.900
Total			$45.778

Sources: Benjamin Holden, "UtiliCorp and Kansas City P&L Agree to Combine in a $1.35 Billion Merger," *Wall Street Journal*, January 23, 1996, p. A3; "Merged Utility Is Named," *New York Times*, December 7, 1995, p. D8; Nicholas Bray and Dawn Blalock, "CSW Bids $2.53 Billion for U.K. Utility; Dallas Power Company Cites Easier Rules in Britain, Tighter Market at Home," *Wall Street Journal*, November 7, 1995, p. A19; Dawn Blalock, "Entergy Intends to Buy CitiPower Australian, Electric Utility," *Wall Street Journal*, November 20, 1995, p. A5; "Three Midwest Utilities Join Rising Trend toward Consolidation," *New York Times*, November 13, 1995, D2, and "IES Industries Stock Climbs 12% on News of MidAmerican Bid," *Wall Street Journal*, August 6, 1996, p. C13; David Cay Johnston, "2 Deals Continue Wave of Mergers in Energy Utilities," *Wall Street Journal*, August 13, 1996, pp. D1, D5; Benjamin A. Holden, "PacifiCorp to Buy Utility in Australia," *Wall Street Journal*, November 16, 1995, p. A3; "Allegheny Power System Inc.: Subsidiary Agrees to Buy Rest of Generating Station," *Wall Street Journal*, November 30, 1995, p. B4; "Two Kansas City Utilities Agree to a Merger," *Los Angeles Times*, January 23, 1996, p. D2; "Kansas City Power Resists Hostile Takeover," *New York Times*, July 10, 1996, p. D4; "UtiliCorp Shareholders Approve Kansas Utility Merger," *New York Times*, August 15, 1996, p. D4; "Judge in Kan-

160

sas Rules on Utility Merger," *New York Times*, August 3, 1996, p. A37; "American Water Works Co.: Unit Buys Water Utility for Total of $262 Million," *Wall Street Journal*, February 20, 1996, p. A6; "Kansas City Power & Light Co.," *Wall Street Journal*, March 16, 1996, B6; "Enron Files Bid to Buy Assets," *Wall Street Journal*, June 18, 1996, p. A8; Monica Langley, "Southern Joins Crowd Bidding for Cajun Electric," *Wall Street Journal*, June 27, 1996, A3; and Agis Salpukis, "Latest Offer for Utility Is $1 Billion," *New York Times*, June 27, 1996, p. D2; Peter Fritsch, "Noble Affiliates to Pay $775 Million for a Unit of New Jersey Utility," *Wall Street Journal*, July 3, 1996, p. B6; Benjamin A. Holden, "Enron Corp. Has Accord to Buy Portland General," *Wall Street Journal*, July 22, 1996,A3; Sullivan Allanna, "Enron Deal Signals Trend in Utilities," *Wall Street Journal*, July 23, 1996, A3; and Allen R. Myerson, "Enron Will Buy Oregon Utility in Deal Valued at $2.1 Billion," *New York Times*, July 23, 1996, p. D1; "Cinergy Says It Is in Talks on a Possible Merger," *New York Times*, August 13, 1996, p. D4; "MidAmerican in Bid for Rival Iowa Utility," *New York Times*, August 6, 1996, D2; "Minnesota Power to Buy Firm," *Wall Street Journal*, August 22, 1996, p. A6; "Southern Union Co.: Suit Filed as Part of Effort to Stop 2 Utilities Mergers," *Wall Street Journal*, August 21, 1996, p. B6; Agis Salpukas, "Ohio Edison to Buy Another Utility for $1.6 Billion," *New York Times*, September 17, 1996 p. D1; "Company News. TransAmerican Waste Drops Its Purchase Effort," *New York Times*, D3; "Laidlaw Inc.: Western Resources to Buy Remaining Stake in ADT," *Wall Street Journal*, March 8, 1996, p. B4; Steven Lipin and Peter Fritsch, "Duke Power Plans to Acquire PanEnergy in Stock Transaction of About $7.7 Billion," *Wall Street Journal*, November 25, 1996, pp. A1, A3; Benjamin A. Holden, "PG&E Agrees to Buy Unit from Valero," *Wall Street Journal*, February 3, 1997, pp. A1, A3; Agis Salpukis, "Enron Utility Merger Approved in Move to Expand Nationwide," *New York Times*, February 27, 1997, D1, D4; Benjamin A. Holden, "PG&E Agrees to Buy Unit from Valero," *Wall Street Journal*, February 3, 1997, pp. A1, A3; Carlos Tejada, "NGC to Acquire Destec for $127 Billion. Natural-Gas Concern Aims to Stake Out Position in Electricity Market" *Wall Street Journal*, February 19, 1997, pp. A1, A2; and Agis Salpukas, "PacifiCorp is Said to Reach Deal to Buy British Utility," *New York Times*, June 12, 1997, pp. D1, D7.

Table 8.2
Pacific Gas & Electric Company
Picks Up Low-Cost U.S. Generation Capacity

Date	Acquisitor	Target	Cost ($ billions)
Nov 1996	Pacific Gas & Electric Co.	Teco Pipeline Co. from TRT Holdings (Corpus Christi)	n.a.
Dec 1996	Pacific Gas & Electric Co.	Energy Source Inc. (Houston)	n.a.
Feb 1997	Pacific Gas & Electric Co.	Valero Energy Corp.	$0.723

Pacific Gas & Electric (San Francisco)	Bond Rating = A PE Ratio = 11.7	Its rates are high. But California regulators will allow it to recover the cost of building expensive nuclear plants.

Sources: Benjamin A. Holden, "PG&E Agrees to Buy Unit from Valero," *Wall Street Journal*, Feb-ruary 3, 1997, pp. A1, A3; Agis Salpukis, "Enron Utility Merger Approved in Move to Expand Nationwide," *New York Times*, February 27, 1997, pp. D1, D4; and Standard & Poor, *Business Week*, ratings and PE ratios taken in December 1996.

Although the California state regulators' gradualist strategy has allowed PG&E time to adjust to competitive markets while paying off their stranded costs at the expense of its consumers, the state has required a quid pro quo. California invited in out-of-state competition for the big company's electricity generation business. This invitation came in October 1966, when California regulators required PG&E to sell four of its gas-fired generating plants for $400 million. This forced divestment was part of California's plan to make California energy markets more competitive. The California state Public Utilities Commission is requiring the state's three major utilities to divest themselves of half of their power generation plants in order to make room for out-of-state competitors.[12]

New York and California Regulators

California's regulatory strategy with PG&E is similar to New York's approach with the Long Island Lighting Company (LILCO). In each case, state regulators have apparently decided to allow the high-cost home-state electricity generator to continue charging above-market rates to state customers. This allows the home-state utility to become more competitive, nationally, by developing the ability to charge lower out-of-state prices. In the case of PG&E, the utility is big enough already to have become competitive out-of-state. It will be interesting to see if LILCO can mimic the second part of PG&E's act by competing out of state too.

Out-of-State Competition

PG&E has also picked up serious competition from Texas-based Enron, another nationwide utility power-marketing giant. In January 1997, Enron announced an alliance with the Northern California Power Agency (NCPA) to supply eleven California municipalities with gas and electricity.[13] Enron's agreement with the NCPA effected 7,000 customers and represented the agency's efforts to provide themselves with at least one other supplier of gas and electricity than the high-cost PG&E. New California state law mandates that some California customers be able to buy power from the cheapest sources, regardless of where the provider is, by the beginning of 1998, and Enron's electricity rates, unlike those of PG&E, are 50 percent lower than the national average.[14] The savings the agency expects from Enron's lower prices would be passed on to the cities served by the non-profit agency and used by those cities to pay for other municipal services like road repair and garbage collection.

The agreement with NCPA represented for Enron another incursion into national markets as a low-cost supplier of gas and electricity. Enron's entry into the California power markets provided a symmetry to the competition between two giant utilities. Both Enron and PG&E could be seen competing outside their home states as low-cost national energy power-marketers. The timing of five of PG&E's strategic moves is shown in Table 8.3.

Enron's Strategy of Energy Convergence

Although PG&E is curently the largest utility, the merger that has set the trend for the current round of utilities mergers in the United States was the $2.1 billion July 1996 merger of Houston-based Enron with the large Oregon utility, Portland General Corporation. The merger was a large one and was one of the first expedited under FERC's new streamline guidelines.[15] There were four other large mergers awaiting FERC adjudication at the time including the gas-to-electricity merger of PG&E and the Valero Energy Corporation.

The key to both Oregon state and FERC approval of the merger was Enron's proving that the merger would result in enough operational synergies to allow them to cut rates for their gas and electricity consumers.[16] Somewhat egotistically, in their initial negotiations with the Oregon Public Utility Commission (OPUC), Texas-based Enron pushed their economic arguments hard, stressing their worldwide experience in negotiations of this kind. But in April 1997, faced by stiff regulatory opposition to their merger, Enron and Portland General doubled their promised rate reductions for their Oregon customers.[17] "Their [Enron's] big mistake was coming in initially with the position that there was no need for a guaranteed rate reduction up front," said Phil Nyegaard, the financial analysis administrator of OPUC.[18]

Enron and Portland amended their merger agreement to give Portland shareholders only 0.9825 shares of Enron stock for each share of Portland stock, instead of making it a 1-for-1 swap. This decreased Enron's purchase offer for Portland from $2.1 billion to $2.06 billion and allowed them to boost their proposed rate reduction from $61 to $141 million dollars, producing an increase in

Table 8.3
Time Line of Strategic Moves by Pacific Gas & Electricity Company

Date	Move
Oct 1996	California regulators require PG&E to sell 4 electricity power plants
Nov 1996	PG&E acquires Teco Pipeline Co. (Corpus Christi)
Dec 1996	PG&E acquires Energy Source Inc. (Houston)
Jan 1997	Enron (Houston) announces an alliance to supply gas and electricity to 11 California cities formerly supplied only by PG&E.
Feb 1997	PG&E acquires Valero Energy Corp.

their rate reduction of $80 million. When a resolution of the negotiations seemed imminent, Nyegaard called OPUC's conflict with Enron and Portland General a "clash of cultures" between Enron, a natural gas and pipeline giant accustomed to competition, and Oregon regulators, whose chief concern was ensuring low rates for customers.[19]

One thing that attracted Enron to Portland General was its low exposure to costly nuclear power projects. This feature made Portland a low-cost generator of electricity. Some analysts thought that the merger rejected Enron's desire to be as prominent in the electricity-generation business as it was in the production, transmission, and marketing of natural gas.[20] But the more obvious reason for acquiring Portland's electricity generating assets was that Enron wanted to be a low-cost national power-marketer of both gas and electricity.[21] It was Enron's innovation to treat these two energy sources as being interchangeable, supplying their customers with whichever of the sources was the lowest cost at any given time. Although the Portland merger gave Enron thousands of new residential customers in Oregon and its surrounding states, in the past, Enron's strategy has proven more successful with its industrial customers who are better able to make the capital investment in equipment which will let them switch easily between gas and electrical power.

The mammoth $7.7 billion gas-to-electricity merger of Duke Power and Pan-Energy announced in November 1996,[22] ratified the idea that a combination of gas and electrical utilities could actually produce a powerful utility.

FERC Rejects a Merger

Although FERC pioneered a new set of rules to speed up utilities mergers, their May 1997 rejection of the $6 billion merger proposed between the Wisconsin Energy Corporation and the Northern States Power Company showed the commission's continued sensitivity to antitrust issues when a merger was a very large one. Each of the corporations in the Wisconsin-Northern States merger was the largest utility in its state, and the utilities planned to merge into a national entity to be called the Primergy Corporation. FERC disallowed the merger finding that it would give Primergy an unwarranted increase in market power. FERC's decision overruled the findings of an administrative law judge that the merger was not anti-competitive.

FERC's new rules for mergers are modeled on the approach of the Federal Trade Commission (FTC) in antitrust cases. The FTC's rules use a checklist approach to merger approvals rather than a legal model which includes the likelihood of lengthy litigation. As long as the utilities' petition for approval can satisfy the commission that the merger has enough synergies to created a high probability of lower consumer rates, the commission is inclined toward approval. This approach is designed to avoid lengthy, adversarial, legal proofs contesting issues of economic efficiency versus market domination under court procedures that are expensive and time consuming.

But FERC's rejection of the Wisconsin-Northern States merger did not give specific reasons for the rejection other than that the commission thought the deal was anti-competitive. "FERC's recent actions make it clear that the commission's policies are still being developed," asserted James J. Howard, chairman, president and CEO of Northern States, Minneapolis. "There is simply no end to this process in sight."[23]

A determining issue in the decision that made this merger proceeding somewhat unique, was that the two partners had abutting territories, and their control of the power transmission lines in their territories made it difficult to see how they could be prevented from freezing out competition. The merger partners had addressed this issue by proposing an independent operator for the power lines of the merged companies.[24] But the equipment-related constraints on energy transmission in the region made the Primergy situation unique, said Merrill Lynch & Company analyst Steven I. Fleishman, "There really isn't another [merger] that's nearly as bad on the wholesale-transmission issue."[25]

Under the new FERC guidelines, mergers between utilities from different regions would draw considerably less scrutiny than combinations of two big electricity companies, such as Northern States and Wisconsin Electricity, who dominate their respective regions and have abutting territories.[26] This approach could make FERC significantly more liberal on national power-marketing mergers between firms that generally do not have abutting territories. The Enron-Portland General combination and the merger between Duke Power and PanEnergy are both examples of this type of merger.

In May 1997, Northern and Wisconsin called off their merger, because of FERC's refusal to approve it. FERC was concerned about the market power that would have been created by merging two such large utilities with abutting market areas. The regulatory refusal was expensive: the two companies had spent $58 million in a fruitless two-year effor to combine into a $6 billion company to be called Primergy Corp.[27] The concern of Primergy's smaller competitors, however, was evident:

Madison Gas and Electricity, a smaller utility based in Wisconsin's capital which has adamantly opposed the Primergy merger, praised FERC's rebuff of the plan as filed as one that will "protect all Wisconsin citizens from the ravages of illegal, monopolistic control of Wisconsin's electricity-energy supply." David C. Mebane Madison's chairman and chief executive officer, said "generation divestiture and a truly independent transmission system operator" are "essential" to protecting energy consumers in the region.[28]

Although the size and contiguity of the merger partners made this merger relatively novel, the court's decision was a flat denunciation of the deal based on its noncompetitive tendency to dominate the upper Midwest power markets. FERC found that Primergy's merger plan did not provide enough access to competitive vendors of power generated elsewhere.[29] These issues are often dealt with by ordering the merger partners to divest themselves of certain units to facilitate competition as was done with PG&E in California. But in this case, the two merging companies were so large relative to their markets that this remedy might not have been effective. With the Wisconsin-Northern States ruling, FERC is simply "saying we cannot approve the merger at this time because it would not be in the public interest. They would dominate the market."[30]

Hostile Versus Friendly Mergers

Western Resources Versus
Kansas City Power & Light

KCP&L is a large generator of electricity, while Western's focus is on transmission and distribution of power and other consumer services.[31] The merger will form a midwestern powerhouse by combining KCP&L's 430,000 customers with Western's 1.2 million gas and electric customers in Kansas and Oklahoma.

Whereas the Wisconsin and the Northern States spent $58 million on the pursuit of a friendly, but unsuccessful merger, Western resources spent $2 billion on a successful nine-month battle[32] to take over the Kansas City Power and Light Company (KCP&L).[33] Western Resources won its battle to take over KCP&L in March 1997, completing what was the first hostile utility merger in the wave of U.S. utilities mergers that started in approximately 1992. With a market value of $2.1 billion, the Western Resources-KCP&L merger was also one of the largest utility mergers as the figures for mergers soared in 1996 to $41.95 billion, up from a mere $2.5 billion in 1993.[34]

The fact that this was the largest successful hostile utilities merger in the United States, proves that hostile takeovers are a feasible technique in an industry where mergers have almost always been friendly.[35] The merger involved a personality conflict between executives John E. Hayes, Jr. of Western Resources and A. Drue Jennings of KCP&L.[36] The acquisitions strategy, however, was guided by David C. Wittig, formerly a mergers and acquisitions specialist for Salomon Brothers, and Wittig's background might account for the adversarial style of the takeover. At the time the KCP&L merger was completed, Western had a pending adversarial $4.3 billion takeover bid for ADT Ltd., the world's largest manufacturer of home security products. ADT, however, mounted an energetic defense, and Tyco International topped Western's bid by offering ADT $5.4 billion in a friendly deal.[37]

Western Resources was formed in 1992 through a merger of the Kansas Gas & Electricity Company (KG&E) and the Kansas Power & Light Company (KP&L) in order to prevent a hostile takeover of KG&E by KCP&L.[38] In January 1996, UtiliCorp United and KCP&L agreed to a $3 billion friendly crosstown merger to fend off Western Resources' initial $1.7 billion hostile bid.[39] But

David C. Wittig, the president of Western Resources, led an active campaign and succeeding in derailing KCP&L's friendly merger with UtiliCorp.[40]

It is easy to see how corporate hubris might soar during this five-year-long series of merger maneuvers, during which the roles of aggressor and target switched until Western Resources acquired the firm had attacked it five years before.[41] But the personal feud between Hayes and Jennings was reported to be a bygone,[42] and Andy Patterson of Mercer Management Consulting commented that the fact that the executives weren't friendly might actually make the hard decisions of the merger easier—there would be less polite beating about the bush.[43]

Managerially Difficult

It is difficult to understand why Western Resources would expend the resources necessary to complete a hostile merger with KCP&L when it is so difficult to get even a friendly merger to work. The basic problem in mergers is to blend different management and operating styles,[44] and the hostile mood of the Western-KCP&L merger could reasonably be expected to make this kind of blend even more difficult than usual. After some mergers, employees feel a sense of disease and say they never feel the same about the new company.[45] One evidence of this lack of blend between Western and KCP&L was the continued use of separate headquarters for each of the two merged companies.

Another peculiar difficulty in utility mergers is the need to secure complex regulatory approvals.[46] There are always negotiations between the merging utilities and their regulatory agencies to arrange the division of the gains from synergies hoped to result from the combination. Although the Western-KCP&L merger was a pre-deregulation merger with no clear guarantees that the customers would get much benefit from the combination, regulators always press for guarantees that any gains from synergies will be shared with consumers. If the regulatory agency determines that the merger is being done to dominate markets, then it is very difficult for them to allow the merger based on any synergies, because the power of the merged companies would leave no competitive reasons for them to release any of their profits in the form of reduced consumer prices.

When these negotiations occur in the courts, it is necessary for utility managers to amass evidence on all of these issues, so the proceedings have a tendency to turn the vision of the firm inward toward industry competitive and regulatory issues, distracting them from the management of efficient utility operations.

The fact that utilities are merging while their markets are becoming more competitive suggests that many of the new mergers are being done to minimize risk, rather than to open new markets and serve customers better. Also negative is the expectation that merging utilities should engage in enough cost-cutting to simultaneously increase the utility's stock value and reduce their power rates. The cost cutting and layoffs necessary to achieve these goals often demoralize the workforce in the utility. Even with the best management, the merger of dissimilar companies often permanently changes the way that employees feel about the merged company.[47]

Are mergers an effective response to deregulation, or are they the utility industry's knee-jerk response to more competition—the response of a herd mentality? Stephen Bucalo, the vice president of the utility and energy practice of A. T. Kearney, a utility consulting firm, thinks that mergers are best used as a last resort strategy.[48] Morton Pierce, who was involved in the Western-KCP&L merger as the head of Dewey Ballantine's mergers and acquisitions group, thinks that all mergers are difficult, although some of them do live up to their promise.[49] To satisfy the financial-economic definition of a "good" merger, the combination should result in significant synergies: a better mix of generating resources leading to lower costs. This is apparently not generally true, though, because a recent survey by Boston-based Mercer Management Consulting showed that of forty-three mergers from 1985 to 1995 only twelve provided superior revenues and operating profit growth.[50]

Western said that their merger with KCP&L would squeeze out $1 billion of costs over a ten-year period without any layoffs. The savings would come from employment attrition and the elimination of duplicate departments, but in a curious twist, Western also said that KCP&L would continue to have its own headquarters in Kansas City—an unusual arrangement.

Western's merger with KCP&L left its product strategy unclear. KCP&L had preferred a merger with UtiliCorp, a natural gas-based utility. The broken merger of KCP&L and UtiliCorp would have put together two companies with similar product lines, resulting in a specialization of skills. This kind of specialization generally leads to an extension of horizontal markets rather than vertical integration. Western, however, was a utility focused on the transmission and distribution of power and other consumer services.[51]

The merger signalled that Western intends to broaden their product line. This broadening was also implied by their bid for ADT and its home security products. A successful acquisition of ADT would have broadened Western's product line so much that it might hae been it might be able to offer one-stop utility shopping. "Western will be transformed into a consumer services company if it succeeds in acquiring ADT," said James L. Dobson, a utility analyst with Donaldson, Lufkin & Jenrette.[52]

UTILITY ACQUISITIONS SPREAD GLOBALLY

U.S. Acquisitions Spread Overseas

Beginning in 1995 the merger activity in the U.S. utilities industry spread to Britain. The merger wave in Britain had begun in December 1994 with Trafalgar House's bid for Northern Electricity. After some initial British-with-British merger skirmishing, the real British merger wave was kicked off in July 1995 by Southern Company's bid for Southwestern Electricity PLC and then, a year later, a bid for Southern Electricity (U.K.) and subsequently a bid for Britain's largest power company, National Power PLC.

The Southern Company Invests
in the United Kingdom

In the period from July 1995 to May 1997, the Atlanta-based Southern Company made an astonishing $15 billion in bids to acquire foreign utilities. Table 8.4 shows these bids. With the acquisition of Southern Electricity PLC and Southwestern Electricity PLC, Southern added approximately $2 billion of British utilities to their portfolio. Over $5 billion of Southern's foreign merger bids were actually completed.

What set the stage for Southern's acquisitions was Britain's program to privatize and deregulate their utilities industry. This program was well under way in 1995 and had already begun to unlock the profit potential in Britain's privatized utilities when the Southern Company began their British utility acquisitions. This profit potential was very attractive to Southern. Unlike many of their competitors, Southern did not choose to merge with nearby utilities and cut costs to get more control over a larger customer base.[53] Instead, Southern looked abroad for generating assets with a much higher rate of return. Southern already provided a large customer base in the United States with low cost power, so it was in a good position to sell power abroad as foreign utility monopolies were stripped away.[54] British deregulation was also responsible for the wave of British utilities mergers which swept the U.K. beginning in late 1994.

The wave of British-with-British utilities mergers began in 1994 with Trafalgar House's bid for Northern Electricity Company. This was followed in August of 1995 by Hanson's bid for Eastern Group, in September by PowerGen's bid for Midlands Electricity, and in December by Welsh Water's bid for South Wales Electricity—over $3 billion in proposed mergers. Two of these mergers failed. Trafalgar House was eventually outbid in a bitter contest with CalEnergy of the United States, and British regulators blocked the PowerGen-Midlands combination because it was deemed anticompetitive. But with Welsh Water's bid for South Wales Electricity, and Wessex Water's $1.31 billion bid for Southwest Water in March 1996, the wave of British mergers was well under way. Table 8.5 shows these British-with-British utilities mergers.

"The voracious" Southern company's British acquisitions, were prompted by Southern's straightforward search for low-cost generating assets with which to sustain their profit margins and return on investment. After a string of U.S. mergers, Southern had experienced difficulty in maintaining average returns on their investments, and had began to look abroad when U.S. regulators disapproved of their plans for a domestic gas merger. After British regulators forced Southern to abandon their acquisition of National Power PLC, Southern began shopping globally and bid $2.7 billion for Consolidated Electricity Power Asia, a Hong Kong-based generating company, and then made a $759 million consortium offer for Berliner Kraft & Licht AG (Bewag), a Berlin electrical power monopoly. All of these mergers offered Southern more profit potential than mergers with

Table 8.4
The Southern Company's
Acquisitions in the U.K., Hong Kong and Berlin

Date	Acquisitor	Target	Bid ($ billions)
Jul 1995	Southern Company (Atlanta)	Southwestern Electricity PLC (U.K.)	$1.650 purchased
Apr 1996	Southern Company (Atlanta)	Southern Electricity PLC (U.K.)	n.a.
Apr 1996	Southern Co. (Atlanta)	National Power PLC (U.K.)	$10.0+ bid dropped May 1996
Oct 1996	Southern Co. (Atlanta)	Consolidated Electricity Power Asia Ltd. (Hong Kong)	$2.700
May 1997	Southern Company (Atlanta)	Berliner Kraft & Licht AG (Bewag) (Berlin)	$0.759

Sources: Matthew C. Quinn, "Southern Co. Pleased with British Venture in Natural Gas," *Atlanta Constitution*, April 30, 1996, p. D2, "British Utility Rejects Offer by Southern," *New York Times*, July 12, 1995, pp. C4, D4; Emory Thomas, Jr., "Southern Company Bids for Firm is Rebuffed," *Wall Street Journal*, July 14, 1995, pp. A3, C22; "Southern in Hostile Bid for British Utility," *New York Times*, July 14, 1995, pp. C16, D16; "Southwestern Electricity PLC," *Wall Street Journal*, July 28, 1995, pp. B2, A6; "British Utility Fights Bid by Southern Company," *New York Times*, July 29, 1995, pp. 19, 37; Lawrence Ingrassia, "Southern Company to Pursue Its Hostile Offer for British Utility as Another Bid Flops," *Wall Street Journal*, August 14, 1995, pp. A7, A6; Richard W. Stevenson, "British Utility to be Acquired by Southern Company" *New York Times*, August 26, 1995, pp. 17, 33; "Southern Company Expansion," *Wall Street Journal*, November 2, 1993, pp. A11, A14; "Southern Seeking British Power Concern," *New York Times*, April 10, 1996, p. D11, "Southern Co.: Plan to Purchase Utility Dropped after U.K. Action," *Wall Street Journal*, May 9, 1996, p. B4; Matthew C. Quinn, "British Utility Seeks Own Merger in an Effort to Follow Southern

Co.," *Atlanta Constitution*, April 13, 1996, p. B2; Matthew C. Quinn, "Southern Co. Stalks Major Acquisition," *Atlanta Constitution*, April 18, 1996, p. B1; Matthew C. Quinn. "British Power Producer Spurns Southern's Overture," *Atlanta Constitution*, April 19, 1996, C1; Matthew Quinn, "Southern Co. Drops Attempt to Buy Second British Utility," *Atlanta Constitution*, May 9, 1996, p. E7; "Southern Co. Unit Getting Role in Asia," *New York Times*, October 10, 1996, p. D4; Benjamin A. Holden and Anita Sharpe, "Southern Co. to Buy Asian Powerhouse, *Wall Street Journal*, October 1, 1996, p. A3; Greg Steinmetz, "Southern Co. Investment in Utility in Berlin May Be Unveiled Today," *Wall Street Journal*, May 13, 1997, pp. A1, A17; and Agis Salpukas, "Big U.S. Utility Spreads Its Reach to Berlin," *New York Times*, May 24, 1997, p. D2.

companies on the somewhat shopped-over list of American utilities could have provided them.

It was Southern Company's bids for British utilities that started the wave of foreign acquisitions of U.K. power companies. Southern Company's first $1.65 billion bid for Southwestern Electricity PLC in July 1995 was successful. In April 1996, the Southern company followed its Southwestern merger with a $10+ billion bid for National Power PLC, Britain's largest electrical power utility, and also with a bid for another British firm, Southern Electric PLC (no relation to the Southern Company). Southern's bid for National Power prompted National to make a defensive bid of $3.79 billion for Southern Electric PLC, hoping that the $14 billion package of National and Southern Electric would be too large for Southern to swallow.

On April 18, 1996, National Power spurned Southern's takeover overture, which the press termed "brash." Although the Southern Company had a market value of $14.7 billion, this was not much more than the value of National Power at $10 billion, and there were doubts that the Atlanta-based utility holding company had the wherewithal to annex Britain's largest power generator.[55]

National Power, rushed to derail Southern's takeover bid by announcing its own merger plans with Southern Electric.[56] National further underscored this defensive strategy by announcing a new, higher offer for Southern Electric, raising the bid to $3.79 billion.[57] National's final bid for Southern Electric was an 8 percent increase over their previous bid in the fall of 1995. At that time, no mergers of electric power producers had been approved in the U.K.. For its part, the Atlanta-based utility reported the success of its previous acquisition (Southwestern Electricity) in Britain's natural gas market.[58]

In May 1996, the Southern Company conceded defeat in their bid for National Power, a week after Britain's Department of Trade and Industry said that it would retain its "golden shares" in the country's electric utilities that give it an effective veto power over mergers and takeovers.[59] The British government wanted National Power to remain independent, so the Southern Company set its sites on more modest expansions in the U.S., Latin America and Caribbean.[60] In June 1996, the Southern Company bid $1 billion for the assets of the bankrupt Cajun Electric Power Coop.(La.); and in July they bid for both Colb Achicura SA (Chilean) and the Jamaica Public Service Company.

Replication: British Utilities Invest Abroad

It was Stephen Littlechild, one of the architects of the plan to deregulate and privatize British utilities, who encouraged British-owned National Power and PowerGen to invest abroad rather than in the U.K.[61] PowerGen's decision to invest abroad was inspired by the British regulator's decision in May 1997, to reverse the findings of the Monopolies and Mergers Commission, and prevent PowerGen from acquiring Midlands Electricity, a large regional generator.[62]

The decision was not surprising, however, because PowerGen was the second largest generator in Britain, and even after being forced to sell some of its domestic power plants, it still retained 21.6 percent of the British electricity mar-

Table 8.5
British-with-British Utilities Mergers

Date	Acquisitor	Target	Bid ($billions)
Dec 1994	Trafalgar House (U.K.)	Northern Electricity (U.K.)	Bid fails.
Aug 1995	Hanson PLC (U.K.)	Eastern Group PLC (U.K.)	n.a.
Sep 1995	PowerGen PLC (U.K.)	Midlands Electricity PLC (U.K.)	Bid blocked, in May (anticompetitive)
Dec 1995	Welsh Water PLC (U.K.)	South Wales Electricity PLC (U.K.)	$1.31
Mar 1996	Wessex Water PLC (U.K.)	Southwest Water PLC (U.K.)	n.a.
Apr 1996	National Power PLC (U.K.)	Southern Electricity PLC (U.K.)	$3.79 defense bid
May 1996	Scottish Power PLC (U.K.)	Southern Water PLC (U.K.)	$1.56 $2.59 wins
May 1996	Southern Electricity PLC (U.K.)	Southern Water PLC (U.K.)	$2.42 outbid.
Dec 1996	Wessex Water PLC (U.K.)	Southwest Water PLC (U.K.)	n.a.

Sources: Lawrence Ingrassia, "Trafalgar House Bids to Take Over Northern Electricity," *Wall Street Journal*, December 20, 1994, A17, A10; "Bid Made for British Utility," *New York Times*, December 20, 1994, pp. C17, D19; Nicholas Bray, "Swiss Bank Role in Trafalgar Bid Sparks Probe," *Wall Street Journal*, January 16, 1995, pp. A7A, B6A; "Northern Electricity Seeks to Reject a $1.9 Billion Bid," *New York Times*, January 24, 1995, p. D3; "Britain Clears Trafalgar House Bid for Utility," *New York Times*, February 15, 1995, C5, D4; "New Offer from Trafalgar," *New York Times*, February 24, 1995, pp. C11, D11; "Trafalgar House Says Bid for Northern Electricity Fails," *Wall Street Journal*, March 13, 1995, p. A13; "U.K. Utility Saga Continues," *Wall Street Journal*, March 13, 1995, A12, A9; Glenn Whitney, "Northern Electricity to Let Trafalgar Make Another Bid," *Wall Street Journal*, March 16, 1995, p. A19; "British Veto Cheaper New Bid for Utility," *New York Times*, March 17, 1995, pp. C4, D4; "British Utility Agrees to Hanson's Cash Bid," *New York Times*, August 1, 1995, C3, D3; Kyle Pope, "Hanson Ventures Into Power Business, Acquiring Eastern Group for $4 Billion," *Wall Street Journal*, August 1, 1995, A3; "Hanson Agrees to Buy Utility," *Los Angeles Times*, August 1, 1995, p. D2; "Britain to Review Utility Acquisitions," *New York Times*, November 24, 1996, p. D12; "PowerGen of Britain

Table 8.5 (continued)

May Bid for Midlands," *New York Times*, September 16, 1995, pp. 18, 32; Tara Parker-Pope, "U.K. Utilities Generate Takeover Frenzy, Latest Bid, PowerGen's for Midlands, Faces Hurdles," *Wall Street Journal*, September 18, 1995, pp. A16, A14; "South Wales Rejects Bid," *New York Times*, December 1, 1995, p. D5; "Welsh Water to Buy South Wales Electricity," *New York Times*, December 5, 1995, p. D4; "Bid for U.K. Water Firm Planned," *Wall Street Journal*, March 8, 1996, p. A8; "Britain to Review," November 24, 1995, D12; Matthew Rose, "U.K. Utility Raises Bid for Rival in Latest Rebuff to a U.S. Suitor," *Wall Street Journal*, April 23, 1996, p. A19; "British Utility Parrying a Bid by Southern," *New York Times*, April 23, 1996, p. D6; "U.K.'s National Power Expected to Bid $4.42 Billion for Southern Electricity PLC," *Wall Street Journal*, October 2, 1995, pp. A12, A6; Tara Parker-Pope, "National Power Joins U.K. Utilities Fray as U.S. Firms, Other Search for Deals," *Wall Street Journal*, October 3, 1995, pp. A15, A17; "Another Giant Merger of British Utilities," *New York Times*, October 3, 1995, pp. C6, D8; Matthew C. Quinn, "British Utility Seeks Own Merger in an Effort to Foil Southern Co.," *Atlanta Constitution*, April 13, 1996, p. B2; Matthew Rose, "U.K. Power Firm Agrees to Acquire Local Water Utility," *Wall Street Journal*, May 30, 1996, p. A11; "Southern Water of Britain Accepts Offer from 2d Bidder," *New York Times*, May 30, 1996, D5; "Takeover Battle Looms for British Utility," *New York Times*, May 31, 1996, p. D4; "British Utility Suitor Withdraws," *Wall Street Journal*, June 21, 1996, p. A8; and "Wessex Water to Bid for Southwest," *New York Times*, March 8, 1996, p. D2.

ket.[63] Were National Power and PowerGen to combine, as they had once discussed, the two firms would control almost half of British power generation.

In the PowerGen decision, the British govenment decided that British firms should not have the market power that they had denied the Southern Company. British free-marketeering regulators remained convinced that levels of industrial concentration as high as those threatened by National Power and PowerGen would hamper the competitiveness of British power markets. So British utilities have been asked to accept lower domestic market shares, forswear the awesome market power at their fingertips in Britain, and, instead, make their acquisitions in competitive markets abroad.[64]

National Power and PowerGen

Having been denied its British merger, PowerGen announced in May 1997 that it intended to invest £421 million to purchase two different plants in the Asia.[65] National Power, the largest power producer in Britain, copied its smaller rival until, together, National Power and PowerGen had poured £4.4 billion into foreign utility investments in 1996 and 1997. National alone invested £1 billion. The investments have gone to Asia, Australia, Europe and the United States.[66] PowerGen's other new overseas investments include a 40 percent stake in an Indonesian coal-fired power station, at a cost of £262 million, and a 30 percent interest in a coal plant in Thailand, which cost £262 million.[67] "Within five years the total output from international projects where we have significant interests will approach the same level as current generation from our U.K. plants," said Deryk King, the managing director of PowerGen group.[68]

The enthusiastic foreign investment of National Power and PowerGen seem to be justified by the fact that foreign power markets, especially those in Asia, are growing faster than those in Britain.[69] This means that both of these companies are satisfied to compensate for their loss of domestic market share by acquiring profitable, fast-growing foreign utilities. But the domestic market shares of the U.K.'s two largest generators is expected to slip even further as more independent power plants start production in Britain.[70]

The foreign investments of National and PowerGen will also serve nicely to diversify their risk under the new windfall profits tax imposed by the Blair government.[71] From the point of view of portfolio diversification, British utility companies may be safer after investing abroad. "International diversification represents a sensible use of the generators' skills," said one analyst. "It is not something their existing investors could easily do themselves."[72]

Whether British utilities have the capacity and skills to manage the riskier foreign investments remains to be seen. In May 1997, the Pakistani government asked to renegotiate their agreement with the 1,292 megawatt Hub power station in which National Power had a 25.7 percent stake. The threat was dispelled but the question of how well qualified the British generator was to compete abroad remained open. The risk of foreign operations appeared to be too much for National Grid Group, which was embarrassed when it was forced to pull out of two international transactions within a week.[73]

Although British utilities are still not certain that the returns from foreign utility investments will prove worth the risk, their profits have been good so far. National Power's after-tax profits on foreign operations in 1997 were £145 million. Although PowerGen expected its current foreign operations to produce about £100 by the year 2001, it made only £12 million overseas in 1997, and £8 million of that was from the sale of assets.[74]

The Adventures of National Power PLC

One of the racier chapters in the recent wave of British-American mergers was that in which National Power PLC, Britain's largest utility, beat off a takeover bid by the Southern Company. National had gotten some fresh experience with utility mergers by going shopping for assets in the U.S. market, where, in July 1993, they had purchased Transco Energy Ventures from Transco Energy Company.

But by April 1996, National had been thoroughly privatized and deregulated, along with all other British utilities, and presented a tempting takeover target. This prompted the Southern Company to make its bid for control of Britain's largest power company. National responded by organizing a $3.79 billion friendly defensive merger with Southern Electricity PLC (U.K.). At that time, the Southern Company was worth $14.7 billion, not much more than National's value of $10 billion, and National hoped that both National and Southern Electricity (U.K.) would be an acquisition too large for the Southern Company to finance.[75] These transactions are shown in Table 8.6.

When Southern made its "brash" bid for National Power, the U.K. had not yet approved any electricity mergers. In April 1996, however, Southern was able to announce the completion of their merger with Southwestern Electricity.[76] But in May, Britain's Department of Trade and Industry decided that it would like the nation's largest utility to remain independent and squelched the Southern Company's bid.[77]

When Southern abandoned its attempt to acquire National Power, National was persuaded to drop its plans to merge with Southern Electricity (U.K.). Now, after American Electric Power's February 1997 acquisition of Yorkshire Electricity, Southern Electricity (U.K.) is the only remaining independent regional electrical power company in Britain.[78] It was thought in 1996 that British regulators might allow another large British utility to take over Southern Electricity as a national champion, and Ed Wallis, the executive chairman, of PowerGen said he would look "very seriously" at buying a regional company if conditions allowed.[79] But with the installation of the Blair government, Britain had decided to direct their larger utilities' investment efforts abroad.

Table 8.6
The Adventures of National Power PLC

Date	Acquisitor	Target	Bid ($billions)
Jul 1993	National Power PLC (U.K.)	Transco Energy Ventures Company (U.S.) from Transco Energy Company	n.a.
July 1995	Southern Co. (U.S.)	Southwestern Electric PLC (U.K.)	$ 1.65 first, successful takeover
Sep 1995	National Power (U.K.)	Southern Electric PLC (U.K.)	$ 3.50 first bid
Apr 1996	Southern Co. (Atlanta)	National Power PLC (U.K.)	$10.00+ bid, dropped May 1996
Apr 1996	National Power PLC (U.K.)	Southern Electricity PLC (U.K.)	$ 3.79 second, defensive bid
Aug 1996	National Power PLC (U.K.), Commonwealth Bank Group (Australia), and PacifiCorp (U.S.) and Destec Energy Inc. (U.S.)	Hazelwood Power Station (Australia)	$ 1.86

Sources: "British Utility to Buy U.S. Power Generation Unit," *New York Times*, July 6 1993, pp. C4, D4; Emory Thomas, Jr. "Southwestern Electricity PLC Accepts Southern's $1.7 Billion Takeover Offer," *Wall Street Journal*, August 28, 1995, p. A5; Matthew C. Quinn, "Southern Co. Pleased with British Venture in Natural Gas," *Atlanta Constitution*, April 30, 1996, p. D2, "British Utility Rejects Offer by Southern," *New York Times*, July 12, 1995, pp. C4, D4; Emory Thomas, Jr., "Southern Company Bids for Firm is Rebuffed," *Wall Street Journal*, July 14, 1995, A3, C22; "Southern in Hostile Bid for British Utility," *New York Times*, July 14, 1995, pp. C16, D16; "Southwestern Electricity PLC," *Wall Street Journal*, July 28, 1995, pp. B2, A6; "British Utility Fights Bid by Southern Company," *New York Times*, July 29, 1995, pp. 19, 37; Lawrence Ingrassia, "Southern Company to Pursue Its Hostile Offer for British Utility as Another Bid Flops," *Wall Street Journal*, August 14, 1995, pp. A7, A6; Richard W. Stevenson, "British Utility to be Acquired by Southern Company" *New York Times*, August 26, 1995, pp. 17, 33; "Southern Seeking British Power Concern," *New York Times*, April 10, 1996, p. D11, "Southern Co.: Plan to Purchase Utility Dropped after U.K. Action," *Wall Street Journal*, May 9, 1996, p. B4; Matthew C. Quinn, "British Utility Seeks Own Merger in an Effort to Follow Southern Co.," *Atlanta Constitution*, April

Table 8.6 (continued)

13, 1996, p. B2; Matthew C. Quinn, "Southern Co. Stalks Major Acquisition," *Atlanta Constitution*, April 18, 1996," p. B1; Matthew C. Quinn. "British Power Producer Spurns Southern's Overture," *Atlanta Constitution*, April 19, 1996, p. C1; Matthew Quinn, "Southern Co. Drops Attempt to Buy Second British Utility," *Atlanta Constitution*, May 9, 1996, p. E7; "Britain to Review Utility Acquisitions," *New York Times*, November 24, 1995, p. D12; Matthew Rose, "U.K. Utility Raises Bid for Rival in Latest Rebuff to a U.S. Suitor," *Wall Street Journal*, April 23, 1996, A19; "British Utility Parrying a Bid by Southern," *New York Times*, April 23, 1996, p. D6; "U.K.'s National Power Expected to Bid $4.42 Billion for Southern Electricity PLC," *Wall Street Journal*, October 2, 1995, pp. A12, A6; Tara Parker-Pope, "National Power Joins U.K. Utilities Fray as U.S. Firms, Other Search for Deals," *Wall Street Journal*, October 3, 1995, pp. A15, A17; "Another Giant Merger of British Utilities," *New York Times*, October 3, 1995, pp. C6, D8; Matthew C. Quinn, "British Utility Seeks Own Merger in an Effort to Foil Southern Co.," *Atlanta Constitution*, April 13, 1996, p. B2; and Benjamin A. Holden, "PacifiCorp and Destec Join Group to Pay $1.86 Billion for a Utility in Australia," *Wall Street Journal*, August 5, 1996, p. A4.

Limits on Britain's Mergers Policy

In June 1997, PacifiCorp made a $5.8 billion bid to buy Britain's Energy Group PLC in the U.S. acquisition of U.K. power assets.[80] The total price of the deal would have been close to $9.6 billion because part of the merger agreement obligated PacifiCorp to assume about $3.8 billion of Energy Group's debt.[81] The amount of financing was so great that PacifiCorp sold its U.S. telephone business to Century Telephone for $1.52 billion to help raise the funds needed to complete the purchase. The bid was part of PacifiCorp's continuing strategy to link up coal, gas, and electricity assets in a way that will allow them to trade contracts for energy equivalents in international power markets.[82] The proposed merger was also the latest U.S. bid for British utilities in which nine American utilities have acquired seven British electricity distribution companies in just two years.[83]

Foreign Investment Equals More Competition

Together, Britain's Trade and Industry secretariat and its Monopolies and Mergers Commission had established a policy of allowing foreign mergers with privatized and deregulated British utilities in the interests of making their utilities industry more competitive and more efficient. The foreign management provided by large modern U.S. utilities promised both more competition with British-owned utilities and lower rates for British consumers.

The corollary to this policy was that British utilities had often been prevented from merging with one another. When their domestic mergers had been blocked as being anticompetitive, British utilities had sought acquisitions abroad. National Power and PowerGen, Britain's first and second largest firms, alone, invested $2.75 billion abroad in 1996 and 1997 as a result of this British antimerger policy. In August 1997, however, Margaret Beckett, Tony Blair's new Secretary of Trade and Industry ordered the review of PacifiCorp's bid for Energy Group PLC. This referral technically ended the merger agreement between the two firms, although the agreement might be renewed.[84] The referral was especially important because it appeared to establish a limit on the number mergers and the degree of market share foreign investors would be allowed in the British utilities industry. From May 1996 to June 1997, U.S. firms have made over $12 billion of bids. The $6 billion bid of PacifiCorp for Energy Group effectively doubled the amount of these bids up to the $12 billion mark. These bids are shown in Table 8.7.

The referral of the PacifiCorp-Energy Group merger was not made on any specific grounds of uncompetitiveness, but rather on the general grounds that the merger might create a segment of the British utilities industry that was beyond governmental control. The loss of control may have simply referred to foreign managerial control, but it might also relate to the type of trading in energy equivalents that PacifiCorp intended. This type of trading would be beyond control in the sense that privatized international energy markets, like markets everywhere, are by their nature uncontrolled if the markets are to achieve an efficient allocation of product at accurate prices. Thus, the cold feet of the new British Labor government in the PacifiCorp merger deal was surprising in light of the bold revamping the utilities industry had undergone during the previous free-marketeering Tory government, and also in light of the recommendations from other British regulators to quickly approve the deal.[85] This may be an instance where it takes governmental policy time to learn about the novel advantages of the new globalized power markets that are emerging.

An even simpler explanation for the decision against the PacifiCorp-Energy Group merger was that the merger was large enough to exceed British regulators' tolerance of excessive market share. Even FERC will not tolerate excessive market share if they feel that the market power that it confers is uncompetitive. So the PacifiCorp-Energy One merger might be said to have been disapproved for the same basic reason as were two other larger mergers, those of Southern-National Power, and National Power-Powergen.

PacifiCorp and Energy Group expressed their disappointment in this shift of policy, but said that they remained committed to the merger and were continuing with the various steps outlined in the deal on the assumption that the merger would eventually take place.[86] Unfortunately for PacifiCorp, one these steps involved a foreign exchange loss of $65 million on currency that had to be ready for payment under the merger agreement. This loss had to be reported on PacifiCorp's quarterly income statement in August 1997.[87]

AEP Acquires Yorkshire Power

Despite their disapproval of the PacifiCorp-Energy Group merger, the British government had not backed down on their free-marketeering policy toward the nation's utilities industry. In February 1997, American Electric Power (AEP) and its partner Public Service Company of Colorado, made a $2.43 billion bid for Yorkshire Electricity.[88] There was virtually no mention of regulatory opposition to this merger although Yorkshire was the last unmerged British regional electrical distributor.

American Electric is based in Columbus, Ohio, and has net profits of $587.4 million on sales of $5.8 billion, providing 132,573 million kilowatt hours of electricity to 7 million customers in seven states.[89] Denver-based PS Colorado is smaller, with an annual income of $190.3 million in 1996 on sales of $2.17 billion. Merrill Lynch advised both of the bidding partners on the Yorkshire acquisition.[90]

Yorkshire Electricity is a well-managed company with some of Britain's lowest cost electrical power and an excellent record of customer service. Colorado would give Yorkshire the additional capability to market gas, while Yorkshire's low-cost production would add to the electric generation earnings power of AEP, earnings of 15 percent on investment, better than AEP could earn in the United States.[91] Although the bid for Yorkshire is a large acquisition for the partners, both AEP and PS Colorado have been shopping globally for high-profit acquisitions. AEP recently acquired the international generating capabilities of Destec International, while PS Colorado merged with Southwestern Public Service

Table 8.7
Other Recent U.S. Utility Acquisitions in the U.K.

Date	Acquisitor	Target	Bid ($billions)
May 1996	General Public Utilities (NJ) and Cinergy (Cincinnati)	Midlands Electricity PLC (U.K.)	$2.00
May 1996	Mystery Suitor (U.S.)	Midlands Electricity PLC (U.K.)	n.a.
Feb 1997	American Electricity Power (Columbus)	Yorkshire Electricity (U.K.)	$2.43
Oct 1996	CalEnergy (Omaha)	Northern Electricity (U.K.)	$1.55
Jun 1997	PacifiCorp (Portland)	Energy Group, PLC (U.K.)	$6.00

Sources: Benjamin Holden, "GPU, Cinergy May Acquire British Utility," *Wall Street Journal*, May 7, 1996, p. A3; "A U.S. Bid Seen Today for Midlands of Britain," *New York Times*, May 7, 1996, p. D7; "U.K. Utility Acknowledges Suitor," *Wall Street Journal*, May 6, 1996, p. A11; Simon Holberton, "Deregulation Has Set U.S. Power Companies Scouting the World for Opportunities," *Financial Times*, February 25, 1997, p. 23; "U.K. Electricity," *Financial Times*, February 25, 1997, p. 23; Agis Salpukas, "Calenegy Offers to Buy British Utility," *New York Times*, October, 19, 1996, pp. D1, D7; "British Utility Lifts Payout in Face of Bid," *New York Times*, December 11, 1996, p. D6; "CalEnergy Bid for Utility is Victorious. British Distributor is Latest Acquisition," *New York Times*, December 25, 1996, pp. D1, D3; James P. Miller Steven Lipin, "CalEnergy Launches Another Hostile Bid," *Wall Street Journal*, July 16, 1997, pp. A1, A3, A4; and Agis Salpukas, "$1.9 Billion Hostile Bid for Utility," *New York Times*, July 16, 1997, pp. D1, D18.

Company, a deal expected to double its generation capability and produce net savings of $77 million over ten years.[92]

The $2.43 billion bid for Yorkshire was thought to be from 15 to 30 percent above its market value. But despite the handsome offer, the price of Yorkshire shares continued selling at a 45 pence discount.[93] One explanation for this was that the stockholders had factored in about a 20 percent probability that the deal would be blocked by the government. This meant that there was an 80 percent chance that the deal would be approved. These optimistic odds were primarily due to the fact that the merger was not nearly as large as the PacifiCorp-Energy Group deal, and there appeared to be no anticompetitive aspects to the merger.

Cracking into European Electricity Monopolies

In May 1977, the Southern Company led a consortium that made a $1.7 billion bid for Berliner Kraft und Licht AG (Bewag), a German electrical utility with a monopoly on Berlin power.[94] The two American companies, Southern and AES, held a 90 percent interest in the bidding consortium which also included Opportunity, a Brazilian investment bank. If the bid was successful, the consortium would take four of eleven seats on the German utility's board of directors and become its operating partner.[95]

If Germany allows the completion of the bid, it would be the first time that a foreign bidder has been allowed to take over a German utility monopoly.[96] Although German utilities are not deregulated, they have some of the highest electricity rates in the world, and this raises the possibility that the merger might set a trend toward the acquisition of European utilities. Although both France and Germany remain wary of competition, the Southern Company plans to increase its presence in these markets, angling for a better position before deregulation takes place and triggers the expected wave of foreign investment.[97]

U.S. Mergers Spread to Australia and Latin America

After World War II, U.S. firms began to engage in the massive amounts of foreign direct investment (FDI) that transformed these firms into multinational corporations. This began with investments in Britain and then spread to other, less culturally congenial parts of the world. Today, mergers and acquisitions account for 16 to 20 percent of all capital budgeting decisions annually, and are considered to be a permanent part of the capital budgeting process rather than a transient phenomenon.[98]

The current spillover of U.S. utility acquisitions into Britain and Australia must now be recognized as just another form of FDI. But, in a world populated by multinational utilities, this FDI could equally well be viewed as simply another capital budgeting decision. Such capital investments might be thought of as the acquisition of ready-made packages of assets instead of a series of piecemeal capital investments that must be organized into a firm. The fact that the package of assets is located abroad is almost incidental in the modern world of global finance.

Global Foreign Direct Investment

As U.S. utilities began to pursue foreign merger targets looking for better utility bargains abroad, the post-World War II pattern of British-first investments does not repeat itself. The FDI acquisitions by U.S. utilities were made in Britain and Australia at the same time. This is yet another fact that underscores that multinational business has become global: when firms look for good foreign investments, they appear to scan the entire globe with their return on investment criteria rather than looking first in any one country.

In 1995 and 1996, U.S. utilities bought approximately $5.0 billion of generating assets in Australia. Table 8.8 presents these acquisitions chronologically, placing Australian acquisitions alongside the contemporaneous U.S. acquisitions of British utilities. What has occurred is an international version of the usual capital budgeting practice of corporations: U.S. utilities acquired foreign utilities with the highest net present values first, working down their ranked list of attractive British and Australian utilities until they had purchased fourteen companies in less than two years. The bulk of British utilities were acquired by Atlanta-based Southern Company. The next $5.0 billion of acquisitions were Australian assets, and were purchased by four other large U.S. electrical utilities: PacifiCorp (Oregon), General Public Utilities Corporation (New Jersey), UtiliCorp United (Kansas), and Destec Energy Inc. (California). The largest of these acquisitions, the purchase of the huge Hazelwood Power Station, was bought by a consortium composed of the Commonwealth Bank Group (Australia), PacifiCorp (Portland) and Destec Energy Inc. (California).

The Southern Company, alone, accounted for approximately 50 percent of all U.S. utility acquisitions in Britain and Australia. The total amount of these investments now stands at approximately $10 billion. Although the Southern Company made $15 billion of bids for British utilities, only $5 billion of these bids were actually completed. By moving decisively to make its offers for British utilities, Southern apparently froze out its four other U.S. competitors. Britain feels that by investing early, Southern got the best prices on U.K. power companies. Research also tends to support this conclusion since these were first time foreign investments for Southern.[99] The response of the competitors was to invest in equally attractive Australian utilities.

The simultaneous acquisition of so many large British and Australian power firms is explained primarily by two phenomenons: first, the privatization and deregulation of British and Australian utilities, and second, oligopolistic reaction in foreign direct utilities investment. The deregulation and privatization of both British and Australian utilities invited larger, more competitive foreign utilities to engage in competition with domestic utilities to improve the efficiency and lower consumer power rates. From the point of view of multinational utilities, the deregulation unlocked new profit potential in foreign utilities, while privatization made this profit available to foreign direct investors.

Table 8.8
U.S. Bids for British and Australian Utilities

Date	Acquirer	Acquisition	Cost ($ billion)
Jul 1995	Southern Company (Atlanta)	Southwestern Electricity PLC	$1.650 purchased
Nov 1995	Entergy (New Orleans)	CitiPower Ltd.	n.a.
Nov 1995	PacifiCorp (Portland)	Powercor (Australia)	$1.600
Nov 1995	General Public Utilities Corp. (NJ)	Solaris Power Ltd. (Melbourne)	$0.356
Nov 1995	UtiliCorp United (Kansas City)[5]	United Energy (Melbourne)	$1.150
Apr 1996	Southern Company (Atlanta)	Southern Electric PLC (U.K.)	n.a.
Apr 1996	Southern Co. (Atlanta)[7]	National Power PLC (U.K.)	$10.000+ bid dropped May 1996
May 1996	General Public Utilities (NJ) and Cinergy (Cincinnati)	Midlands Electricity PLC (U.K.)	$2.000 bid
May 1996	Mystery Suitor (U.S.)	Midlands Electricity PLC (U.K.)	n.a.
May 1996	Southern Electric PLC (U.K.)	Southern Water PLC (U.K.)	$2.420 outbid
Aug 1996	National Power PLC (U.K.), Commonwealth Bank Group (Australia), and PacifiCorp (Portland) and Destec Energy Inc. (U.S.)[11]	Hazelwood Power Station (Australia)	$1.860
Oct 1996	Southern Co. (Atlanta)	Consolidated Electric Power Asia Ltd. (U.K.)	$2.700
Oct 1996	CalEnergy (Omaha)	Northern Electric (U.K.)	$1.550
Nov 1996	Dominion Resources Inc.	East Midlands Electricity PLC (U.K.)	$2.150
Jun 1997	PacifiCorp (Portland)	Energy Group, PLC (U.K.)	$6.000
Total			$21.436

Sources: Matthew C. Quinn, "Southern Co. Pleased with British Venture in Natural Gas," *Atlanta Constitution*, April 30, 1996, p. D2; "British Utility Rejects Offer by Southern," *New York Times*, July 12, 1995, pp. C4, D4; Emory Thomas, Jr., "Southern Company Bids for Firm is Rebuffed," *Wall Street Journal*, July 14, 1995, pp. A3, C22; "Southern in Hostile Bid for British Utility," *New York Times*, July 14, 1995, pp. C16, D16; "Southwestern Electricity PLC," *Wall Street Journal*, July 28, 1995, pp. B2, A6; "British Utility Fights Bid by Southern Company," *New York Times*, July 29, 1995, pp. 19, 37; Lawrence Ingrassia, "Southern Company to Pursue Its Hostile Offer for British Utility as Another Bid Flops," *Wall Street Journal*, August 14, 1995, pp. A7, A6; Richard W. Stevenson, "British Utility to be Acquired by Southern Company," *New York Times*, August 26, 1995, pp. 17, 33; Dawn Blalock, "Entergy Intends to Buy CitiPower Australian, Electric Utility," *Wall Street Journal*, November 20, 1995, p. A5; Benjamin A. Holden, "PacifiCorp to Buy Utility in Australia," *Wall Street Journal*, August 13, 1996, pp. D1, D5; Benjamin A. Holden, "2 Deals Continue Wave of Mergers in Energy Utilities," *Wall Street Journal*, November 16, 1995, p. A3; "Southern Company Expansion," *Wall Street Journal*, November 2, 1993, pp. A11, A14; "Southern Seeking British Power Concern," *New York Times*, April 10, 1996, p. D11; Southern Co.: Plan to Purchase Utility Dropped after U.K. Action," *Wall Street Journal*, May 9, 1996, p. B4; Matthew C. Quinn, "British Utility Seeks Own Merger in an Effort to Follow Southern Co.," *Atlanta Constitution*, April 13, 1996, p. B2; Matthew C. Quinn, "Southern Co. Stalks Major Acquisition," *Atlanta Constitution*, April 18, 1996," p. B1; Matthew C. Quinn, "British Power Producer Spurns Southern's Overture," *Atlanta Constitution*, April 19, 1996, p. C1; Matthew Quinn, "Southern Co. Drops Attempt to Buy Second British Utility," *Atlanta Constitution*, May 9, 1996, p. E7; Benjamin Holden, "GPU, Cinergy May Acquire British Utility," *Wall Street Journal*, May 7, 1996, p. A3; "A U.S. Bid Seen Today for Midlands of Britain," *New York Times*, May 7, 1996, p. D7; "U.K. Utility Acknowledges Suitor," *Wall Street Journal*, May 6, 1996, p. A11; Matthew Rose, "U.K. Power Firm Agrees to Acquire Local Water Utility," *Wall Street Journal*, May 30, 1996, p. A11; "Southern Water of Britain Accepts Offer from 2d Bidder," *New York Times*, May 30, 1996, p. D5; "Takeover Battle Looms for British Utility," *New York Times*, May 31, 1996, p. D4; "British Utility Suitor Withdraws," *Wall Street Journal*, June 21, 1996, p. A8; Benjamin A. Holden, "PacifiCorp and Destec Join Group to Pay $1.86 Billion for a Utility in Australia," *Wall Street Journal*, August 5, 1996, p. A4;

Table 8.8 (continued)

"Southern Co. Unit Getting Role in Asia," *New York Times*, October 10, 1996, p. D4: Benjamin A. Holden and Anita Sharpe, "Southern Co. to Buy Asian Powerhouse, *Wall Street Journal*, October 1, 1996, p. A3; Agis Salpukas, "Calenergy Offers to Buy British Utility," *New York Times*, October, 19, 1996, pp. D1, D7; "British Utility Lifts Payout in Face of Bid," *New York Times*, December 11, 1996, p. D6; "CalEnergy Bid for Utility is Victorious. British Distributor is Latest Acquisition," *New York Times*, December 25, 1996, pp. D1, D3; "Dominion Agrees to Buy Power Utility," *New York Times*, November 14, 1996, pp. D1, D4; James P. Miller and Steven Lipin, "CalEnergy Launches Another Hostile Bid," *Wall Street Journal*, July 16, 1997, pp. A1, A3, A4; and Agis Salpukas, "$1.9 Billion Hostile Bid for Utility," *New York Times*, July 16, 1997, pp. D1, D18.

Oligopolistic Reaction in FDI

Oligopolistic reaction, however, decrees that herd instinct rules among the leading market-share rivals in any industry governed by oligopolistic competition.[100] In these industries, investors mimic the competitive foreign investment moves of their rivals to avoid losing critical market share and being frozen out of the game.[101] Oligopolistic reaction in FDI leads to large quick investments being made without much premeditation once the lead firm has set the investment pattern.[102] This behavior describes the pattern of the Southern Company's investments in Britain and the subsequent merger scramble for British and Australian market share.

What distinguishes recent FDI from the post-World War II FDI in Europe is that the U.S. utilities industry was not nearly as industrially concentrated as they are now. Although the utilities competition in most states was either oligopolistic or monopolistic before deregulation, as deregulation approached, that began to change. After deregulation, national competition was possible, so state markets were no longer relevant. The ability to compete in national markets was much more important because any state utility could be attacked in its home state territory by a national power-marketing firm.

FERC allowed rapid mergers among the large utilities and this created the national power marketing firms and thus the national utilities industry. After merging, these firms amassed market share quickly by attacking state markets. This new market share, although impressive in dollar terms, was insignificant nationally, because the new national market was so large.

What Causes Oligopolistic Behavior

Oligopolistic reaction in foreign direct utility investment occurred among the top tier of large U.S. utilities. To preserve the access to capital markets necessary to let them invest and continue to defend themselves against their power-marketing competitors in the United States, they needed to maintain and increase their return on investment. To do this they made many large, quick foreign investments in lucrative deregulated utilities through the device of more than $10 billion of mergers and acquisitions in the U.K., Australia, Asia, and Germany.

This FDI follows fairly closely the classical pattern of rivalry called oligopolistic reaction as described by Frederick Knickerbocker[103] and Edward Flowers.[104] One difference in the utilities investment that has swept outward from the United States to Britain and Australia is that this investment involved global scanning and was made simultaneously in two most appropriate countries. The investment following World War II was more tentative, it was made by building factories rather than through mergers and acquisitions, and the investmests seemed to have been made first in countries that were perhaps culturally more congenial to the United States.

European Rails

European railroads are undergoing a restructuring similar to the program of privatization and deregulation that the British government imposed on its utilities industry. European governments are slashing their budgets, cutting millions of dollars of railroad subsidies, and selling off state-owned railroad franchises to provide cash and invite new competitors into an industry where feather-bedding has been common and railroad jobs were often considered workfare.[105] According to the World Bank, nine nations have already privatized their railroads, while sixteen more are expected to follow, including Australia and Pakistan.[106]

When a European nation's ratio of debt is high relative to its gross national product (GNP), it is likely to be suffering from the kind of budgetary deficits that subsidized railroads exacerbate. Table 8.9 shows the debt of European countries as a percent of gross domestic product (GDP), this table indicates that France and Germany have very high ratios of debt to GNP, and should be especially interested in privatizing their state-owned railways if the national debt is any consideration.

More Efficient Rail Operations

Like Britain, European countries have assumed that competitive, privately-owned railroads will provide better, cheaper service than state-owned railroads. Most of this privatization and deregulation is occurring not in Europe's modern, high-speed commuter railroads, but in the freight lines that are becoming an increasingly important part of European industrial infrastructure.

It is the inefficiency of European freight rail operations that provides the profit potential which has been luring foreign investment into European and Latin America. U.S. rail companies see obvious operational problems that they could correct if they owned the foreign operations. Correcting these problems would decrease costs, increase profits, and enhance share value. This makes the incentives for U.S. firms to buy European rails even simpler than their reasons for acquiring British, Australian, and German utilities. Some of the problems that American rail firms see are:[107]

Britain
- Twice as many employees as are needed
- As many as 50 percent of the locomotives do not work reliably
- Some freight railroads use thirty-year-old computer-routing systems

Europe
- Different rail gauges are used in different countries
- Crews and locomotives are often changed at the borders
- Low-hanging wires prevent the use of double-stacked trains

Argentina
- Railroad telephone communication networks dating to the 1920s
- Only one train runs a day
- Trains are signalled by pulling levers by hand
- As many as 50 percent of the locomotives do not work reliably

Paraguay
- Still uses steam-fired locomotives
- Telegraphs are still used for signaling systems

These inefficiencies mean that European and Latin American railroads are being sold cheaply. American railroads intend to make a profit with their railroad management contracts or acquisitions through managerial innovations to eliminate the problems and improve operational efficiency. Some of these innovations are:[108]

- Replace outdated equipment
- Lay off as much as 50 percent of the workforce
- Implement innovative, intermodal, road-rail freight delivery systems
- Use modern computers and routing software
- Install modern communications equipment

Modernized Rail Systems

An important part of the managerial innovations that are reducing European rail costs are modern, containerized intermodal (road-rail) rail shuttles. In 1994 Sea-Land set up a joint venture with German and Netherlands partners to run a new shuttle connecting Rotterdam to Germany's industrial regions. The partners were able to run the line at rates from 10 to 20 percent cheaper than competitive railroads. The shuttle used containerized freight loaded at the huge Rotterdam docks and delivered it in one shot to its destination without having to shuffle rail cars. This operation was so successful that in April 1997, Sea-Land's parent, CSX, entered a joint venture with the Dutch company NS Cargo and the freight division of Deutsche Bahn to run a Hamburg-to-Milan freight shuttle. In January 1997, the partners entered another agreement for a similar shuttle from Rotterdam to Munich.[109]

The new containerized, intermodal shipping rates are competitive with those of the trucking industry and less subject to weather and traffic jams. Benetton has found this form of shipping so good that it has changed its shipping "philosophy" and now ships 52 percent of its 700 containers a day from northern Italy to 120 countries via containerized, intermodal shipping.[110]

Scope for Expansion

In Europe, railroads carry just 15 percent of the freight. In Latin America the percentage is even lower—8 to 12 percent—whereas in the United States over 40 percent of freight goes over the rails.[111] This potential for expansion makes an investment in a privatized European or Latin American railroad look attractive. A U.S. railroad might reasonably expect to improve the acquired railroad's operational efficiency by 30 percent and then expand their operations an additional 30 percent and make a good profit doing it. Table 8.9 shows the rail share of European countries as a percentage of the freight market and allows a comparison to the same percentage in the United States.

Table 8.9
Unproductive European Rail Transportation

Country	Debt in $Billions	Debt as % of GDP*	Revenues per Employee[1]	Rail Share % of the Freight Market[2]
Britain	7.96	0.7	35,880	8.8
France	34.48	2.6	43,080	23.6
Italy	50.48	4.9	19,440	12.6
Germany[3]	6.95	0.3	31,080	20.4
Spain	9.77	2.0	25,560	7.2
United States	12.13	0.2	155,000	40.6

Notes: [1] Figures are for end of 1994, the latest available, except the United States, which is for 1995.

[2] For 1993, except for France (1994), and the United States (1995).

[3] Recapitalized in 1993.

Source: John Tagliabue, "Europe's Rail Freight on New Path. Revamping with Technology, Competition and U.S. Aid," *New York Times,* February 20, 1997, p. D15.

Table 8.10

U.S. Rail Companies Buy into Overseas Rails

United States: Acquiring Company	Country
RailTex Inc.	Brazil
Wisconsin Central	Britain
Anacostia & Pacific Co.	Chile
Union Pacific Corp.	Mexico
Wisconsin Central	New Zealand
Kansas City Southern Industries Inc.	Panama
CSX Corp.	Germany
CSX Corp.	The Netherlands

Source: Anna Wilde Mathews and Jonathan Friedland, "U.S. Railroads Are Making Tracks Overseas," *Wall Street Journal*, May 21, 1997, p. 4.

U.S. Railroads Buy into Overseas Rails

The opportunity for double-barreled profitability from increased operational efficiency and expanded operations is luring U.S. investment into overseas rails. Table 8.10 lists the recent foreign railroad acquisitions of U.S. rail companies. Table 8.11 gives the details of nine recent U.S.-foreign rail mergers.

The profit potential in overseas rails is not only high, but very risky. "There's terrific opportunity, but there are huge risks" for the U.S. railroads, says Anthony Hatch of NatWest Securities. "These (foreign) companies haven't been thought of as profit-making operations. They're often a form of workfare."[112] When some rail workforces have been reduced, the remaining workers are so demoralized, they almost quit working, making the railroads undependable.[113] Layoffs are virtually inevitable, however. Table 8.9 shows that revenue per employee in the United States is from 360 to 815 percent greater than in Europe.

The nonstandard nature of European rail equipment and rail gauges is also a problem that is neither quickly nor cheaply solved. "A Dutch locomotive gets to the German border and it doesn't work anymore," said Gordon Mott, a CSX official.[114] Although Britain invented rail locomotives, their lines are such a patchwork of different gauges, that they might appear to be a collection of hobby railroads.

These problems make it difficult to run a modern European railway operation across national borders. In Europe, there are three widths of track and seven electrical standards, topped off by a jumble of signal systems, and national agreements on standards are years away. It is often necessary to unload and reload freight at border bottlenecks to accommodate the nonstandard track gauges.[115]

Despite the problems, U.S. management efforts seem to be going well in foreign rails. Wisconsin Central increased the profits from a New Zealand railroad acquisition 154 percent to $37.4 million for the year ending June 30, 1994,[116]

Table 8.11
Mergers and Acquisitions

Date	Takeover	Target
1994	Sea-Land	European Rail Shuttle (German and Netherlands partners) service between Rotterdam and industrial regions of Germany and Northern Italy.
Dec 1996	Wisconsin Central Transportation Corp.	Rail Freight Distribution (handles the British end of English Channel tunnel rail transport)
Dec 1996	Wisconsin Central Transportation Corp.	English Railway
Dec 1996	Wisconsin Central Transportation Corp.	Scottish Railway
Dec 1996	Wisconsin Central Transportation Corp.	Welsh Railway
1996	CSX (Sea-Land's parent) a Rotterdam shuttle to industrial Germany and Northern Italy.	NDX [a JV with NS Cargo (Dutch) and the freight division of Deutsche Bahn]
1996	Kombiwerke (trucking cooperative) a shuttle between German industrial cities, Milan, and Barcelona.	Partners
Jan 1997	CSX (Sea-Land's parent) a Rotterdam-to-Munich freight shuttle.	NDX
Apr 1997	CSX (Sea-Land's parent) a Hamburg-to-Milan freight shuttle.	NDX

Source: John Tagliabue, "Europe's Rail Freight on New Path. Revamping with Technology, Competition and U.S. Aid," *New York Times*, February 20, 1997, p. D15.

and Kansas City Southern plans to triple the freight revenue from their Northeast Railway line in Mexico to $941 million in five years.[117]

These profits are being produced by more efficient operations. In Brazil, Rail-Tex, Inc., a Texas rail company, has increased the number of locomotives operating daily from 160 to 200 and cut derailments and injuries by more than 50 percent.[118] In Mexico, in some cases, U.S. management has doubled the average train speed which started out at 32 kilometers per hour, half that of the average U.S. train.[119] Mexico U.S. managed railroads generally run trains carrying 66 cars rather than the 39 cars that Mexican managed railroads run.[120]

NOTES

1. R. Charles Moyer, "The Future of Electricity," *Business Economics*, October 1996, 31(4): 13-18.

2. *Ibid.*

3. Charles H. Studness, "Converging Markets: The First Real Electric/Gas Merger," *Public Utilities Fortnightly*, October 1, 1996, 134(18): 21-25; and Gary McWilliams, "Enron's Pipeline into the Future," *Business Week*, December 2, 1996, p. 82.

4. Peter Coy and Gary McWilliams, "Electricity: The Power Shift Ahead," *Business Week*, December 2, 1996, 78-82; Charles V. Bagli, "A New Breed of Power Broker; Bankers are Changing the Shape of the Utility Industry," *New York Times*, December 10, 1996, pp. D1, D9; and Agis Salpukas, "Pacific Gas and Electric to Sell 4 Power Plants," *New York Times*, October 23, 1996, p. D4.

5. James P. Miller and Steven Lipin, "CalEnergy to Buy Back Stake Held by Peter Kiewit Sons' for $1.16 Billion," *Wall Street Journal*, September 5, 1997, p. B9; "Entergy Will Pay $2.1 Billion for a Utility, London Electricity," *New York Times*, December 19, 1996, pp. D1, D5; Agis Salpukas, "Calenegy Offers to Buy British Utility," *New York Times*, October, 19, 1996, pp. D1, D7.

6. Coy and McWilliams, "Electricity: the Power Shift," pp, 78-82.

7. Paul Feine, "Making the Mix," *Energy Economist*, July 1996, n177, pp. 8-10, and Ann Monroe, "Big Deals Advance Fusion of Gas/Power Industries," *Investment Dealer's Digest*, October 7, 1996, 62(41): 22.

8. Coy and McWilliams, "Electricity: the Power Shift," pp. 78-82.

9. Agis Salpukas, "Entergy Agrees to Manage Troubled Maine Nuclear Plant," *New York Times*, January 9, 1997, pp. D1, D6; and Susan Jackson, "The Millstones Around NU's Neck," *Business Week*, December 2, 1996, p. 80.

10. Coy and McWilliams, "Electricity: The Power Shift," pp. 78-82, and Salpukas, "Entergy Agrees," pp. D1, D6.

11. Robert J. Michaels, "Electric Utilities Mergers: The Answer or the Question?" *Public Utilities Fortnightly*, January 1, 1996, pp. 20-23.

12. Salpukas, "Pacific Gas and Electric to Sell," p. D4.

13. Peter Fritsch, "Enron to Unveil Energy Alliance Inolving 11 Cities in California," *Wall Street Journal*, January 15, 1997, pp. A1, B4.

14. *Ibid.*

15. Agis Salpukis, "Enron Utility Merger Approved in Move to Expand Nationwide," *New York Times*, February 27, 1997, pp. D1, D4.

16. *Ibid.*

17. Terzah Ewing, "Enron and Portland General Reduce Stock-Swap Ration and Boost Rate Cuts", *Wall Street Journal*, April 15,1997, p. A6.

18. *Ibid.*

19. Benjamin A. Holden, "Enron Corp. Has Accord to Buy Portland General," *Wall Street Journal*, July 22, 1996, p. A3; Sullivan Allanna, "Enron Deal Signals Trend in Utilities," *Wall Street Journal*, July 23, 1996, p. A3; and Allen R. Myerson, "Enron Will Buy Oregon Utility in Deal Valued at $2.1 Billion," *New York Times*, July 23, 1996, p. D1.

20. *Ibid.*

21. *Ibid.*

22. Steven Lipin and Peter Fritsch, "Duke Power Plans to Acquire PanEnergy in Stock Transaction of About $7.7 Billion," *Wall Street Journal*, November 25, 1996, pp. A1, A3.

23. James P. Miller and Benjamin A. Holden, "Northern States Power and Wisconsin Energy Called Off Their Merger, Citing the Unexpected Regulatory Complications They've Encountered," *Wall Street Journal*, May 19, 1997, pp. A1, A4.

24. James P. Miller and Benjamin A. Holden, "Wisconsin Energy Northern States Pact Rejected," *Wall Street Journal*, May 15, 1997, p. A1,

25. Miller and Holden, "Northern States Power," pp. A1, A4.

26. *Ibid.*

27. Miller and Holden, "Northern States Power," pp. A1, A4.

28. James P. Miller and Benjamin A. Holden, "Wisconsin Energy Northern States Pact Rejected," *Wall Street Journal*, May 15, 1997, p. A1.

29. Miller and Holden, "Northern States Power," p. A1, A4.

30. Miller and Holden, "Wisconsin Energy," p. A1, A5.

31. Charles V. Bagli, "$2 Billion Deal Expected For Utility, Kansas City Power Said To Agree to Takeover," New York Times, February 7, 1997, pp. D1, D5.

32. Agis Salpukas, "Now Comes Hard Part in Utility Deal; Mergers May Not be the Panacea the Industry is Seeking," *New York Times*, February 10, 1997, p. D2.

33. Charles V. Bagli, "ADT and Tyco Plan to Merge in $5.4 Billion Stock Swap," *New York Times*, March 18, 1997, pp. D1, D21.

34. Statistic from Securities Data Company cited in Bagli, "ADT and Tyco," pp. D1, D21.

35. Bagli, "$2 Billion Deal," pp. D1, D5.

36. Salpukas, "Now Comes Hard Part," p. D2.

37. Bagli, "ADT and Tyco," pp. D1, D21.

38. "Kansas City Power & Light Co.," *Wall Street Journal*, March 16, 1996, p. B6.

39. Anonymous, "Two Kansas City Utilities Agree to a Merger," *Los Angeles Times*, January 23, 1996, p. D2.

40. Benjamin Holden, "Utility Chief Tests Strategy for New Era," *Wall Street Journal*, May 29, 1996, p. B1.

41. "Kansas City Power & Light Co.," p. B6.

42. Bagli, "$2 Billion Deal Expected," pp. D1, D5.

43. Salpukas, "Now Comes Hard Part," p. D2.

44. *Ibid.*

45. Peter Truell, "In Financial Arena, Merger is the Game; Major Players Seek Lift," *New York Times*, July 5, 1997, pp. A35, A38.

46. *Ibid.*

47. *Ibid.*

48. *Ibid.*

49. *Ibid.*

50. *Ibid.*

51. Bagli, "$2 Billion Deal Expected," pp. D1, D5.

52. Bagli, "ADT and Tyco," pp. D1, D21.

53. Agis Salpukas, "Big U.S. Utility Spreads Its Reach to Berlin," *New York Times*, May 24, 1997, p. D2.

54. *Ibid.*

55. Matthew C. Quinn, "British Power Producer Spurns Southern's Overture," *Atlanta Constitution*, April 19, 1996, p. C1.

56. Matthew C. Quinn, "British Utility Seeks Own Merger in an Effort to Foil Southern Company," *Atlanta Constitution*, April 13, 1996, p. B2; and Matthew C. Quinn, "British Utility Seeks Own Merger in an Effort to Foil Southern Company," *Atlanta Constitution*, April 13, 1996, p. B2.

57. Matthew Rose, "UK Utility Raises Bid for Rival in Latest Rebuff to a U.S. Suitor," *Wall Street Journal*, April 23, 1996, p. A19.

58. Matthew C. Quinn, "Southern Company Pleased with British Venture in Natural Gas," *Atlanta Constitution*, April 30, 1996, p. D2.

59. "Southern Company: Plan to Purchase Utility Dropped After UK Action," *Wall Street Journal*, May 9, 1996, p. B4.

60. Matthew C. Quinn, "Southern Company Drops Attempt to Buy Second British Utility," *Atlanta Constitution*, May 9, 1996, p. E7.

61. Simon Holberton, "Generators Plug in Abroad," *Financial Times*, May 26, 1997, 17.

62. Christine Buckley, "PowerGen Spends £421M on Stakes in Two Far East Plants," *Financial Times*, May 23, 1997, pp. 26, 28.

63. *Ibid.*

64. Holberton, "Generators Plug in," p. 17.

65. Buckley, "PowerGen Spends £421M," pp. 26, 28.

66. Holberton, "Generators Plug in," p. 17.

67. Buckley, "PowerGen Spends £421M," pp. 26, 28.

68. Holberton, "Generators Plug in," p. 17.

69. *Ibid.*

70. Buckley, "PowerGen Spends £421M," pp. 26, 28.

71. *Ibid.*

72. Holberton, "Generators Plug in," p. 17.

73. *Ibid.*

74. *Ibid.*

75. Quinn, "British Utility Seeks," p. B2.

76. Quinn, "Southern Company Pleased," p. D2.

77. "Southern Company: Plan to Purchase," p. B4.

78. "UK Electricity," *Financial Times*, February 25, 1997, p. 23.

79. Buckley, "PowerGen Spends £421M," pp. 26, 28.

80. Benjamin A. Holden, "PacifiCorp is Likely to Finance Purchase in U.K. With Sale of Telecom Business," *Wall Street Journal*, June 17, 1997, p. B4.

81. Agis Salpukas, "PacifiCorp is Said to Reach Deal to Buy British Utility," *New York Times*, June 12, 1997, pp. D1, D7.

82. Benjamin A. Holden, "PacifiCorp Pursues Energy-Swap Plans," *Wall Street Journal*, June 16, 1997, pp. A1, B4.

83. Salpukas, "PacifiCorp is Said," pp. D1, D7.

84. Benjamin A. Holden and Helene Cooper, "Britain to Conduct Antitrust Review of PacifiCorp Deal," *Wall Street Journal*, August 4, 1997, pp. A1, B5.

85. *Ibid.*

86. *Ibid.*

87. Benjamin A. Holden, "PacifiCorp to Take $65 Million Charge on Currency Deals," *Wall Street Journal*, August 13, 1997, pp. A1, A6.

88. Simon Holberton, "Deregulation Has Set US Power Companies Scouting the World for Opportunities," *Financial Times*, February 25, 1997, p. 23.

89. *Ibid.*

90. Anonymous, "Merrill Lynch Takes Dual Role in Electricity Deal," *Financial Times*, February 25, 1997, p. 23.

91. Holberton, "Deregulation Has Set," p. 23.

92. *Ibid.*

93. "UK Electricity," p. 23.

94. Greg Steinmetz, "Southern Co. Investment in Utility in Berlin May be Unveiled Today," *Wall Street Journal*, May 13, 1997, pp. A1, A17.

95. Jonathan Wheatley, "US Utilities Take Cemig Stake," *Financial Times*, May 29, 1997, p. 6.

96. Salpukas, "Big U.S. Utility Spreads," *New York Times*, p. D2.

97. *Ibid.*

98. Thomas E. Copeland and J. Fred Weston, "Mergers, Restructuring, and Corporate Control: Theory," chap. 19 in *Financial Theory and Corporate Policy*, 3d ed. (Reading: Addison-Wesley Publishing Company, 1988): 680-682.

99. John Doukas, "The Effect of Corporate Multinationalism on Shareholders' Wealth: Evidence from International Acquisitions," *Journal of Finance*, December 1988, 43(5): 1181-1175.

100. Frederick T. Knickerbocker, *Oligopolistic Reaction and Multinational Enterprise* (Boston: Division of Research, Graduate School of Business Administration, Harvard University, 1973).

101. Edward B.Flowers, "Oligopolistic Reaction in European and Canadian Direct Investment in the United States," *Journal of International Business Studies*, Fall/ Winter 1976, 7: 43-55.

102. Flowers, "Oligopolistic Reaction in European," pp. 43-55, and Knickerbocker, *Oligopolistic Reaction.*

103. Knickerbocker, *Oligopolistic Reaction.*

104. Flowers, "Oligopolistic Reaction in European," pp. 43-55.

105. Anna Wilde Mathews and Jonathan Friedland, "U.S. Railroads are Making Tracks Overseas," *Wall Street Journal*, May 21, 1997, p. 4.

106. *Ibid.*

107. *Ibid.*

108. *Ibid.*

109. John Tagliabue, "Europe's Rail Freight on New Path; Revamping with Technology, Competition and U.S. Aid," *New York Times*, February 20, 1997, pp. D1, D15.

110. *Ibid.*

111. Mathews and Friedland, "U.S. Railroads," p. 4.

112. *Ibid.*

113. *Ibid.*

114. *Ibid.*

115. Tagliabue, "Europe's Rail Freight," pp. D1, D15.

116. Mathews and Friedland, "U.S. Railroads," p. 4.

117. *Ibid.*

118. *Ibid.*

119. *Ibid.*

120. *Ibid*

Chapter 9

The Globalization of Energy and Utilities

Since 1972 there have been approximately $70 billion of mergers in the U.S. utilities industry. Depending on how many of these bids are finally completed, approximately $37 billion, a full 53 percent, will be combinations with overseas firms, primarily utilities in Britain and Australia. Seeking higher rates of return, U.S. utilities have leapt at the opportunities in the recently privatized and deregulated utilities of Britain and Australia.

This wave of mergers was initiated by state and federal agency efforts to deregulate the production of gas and electricity. The goal of this deregulation was to produce competitive markets in which the price of coal, gas and electricity would be bid down from their current monopolistic levels to levels more congenial to energy consumers. With competitive markets coming, firms began to merge, first to gain economies of scale and lower costs, and then also to combine coal, gas, and electrical operations so that they might effectively market the cheapest source of power nationwide—to become national power marketing firms.

The mergers creating these firms took place mostly in the coal, gas, and electrical sectors industry at a time when a generalized convergence of all of the forms of energy purchased from power companies was occurring. What caused this energy convergence was the adoption of new production technologies allowing increasingly efficient cross conversions of almost all the major forms of energy. In the United States, coal-fired electrical generation was still the cheapest, but the new techniques of gas-fired generation combined with dramatically increasing world supplies of natural gas, dictated that gas-fired generation would set the sustainable price of electricity for the next ten years.

It is an understatement to say that these forces are restructuring first the U.S. utilities industry, and then the global power industry. Some of the changes which are occurring are:

- FERC is leading state and federal agencies in deregulating power supply industries in order to reduce energy costs by as much as a third and save the economy as much as $50 billion a year in energy costs.[1]
- In order to secure regulatory approval, merger plans often provide that transmission lines be put under independent operators in order to provide power suppliers competitive access to customers to prevent the merger partners from gaining undue market power.[2]
- Cities and towns have been encouraged to condemn municipal power lines, and set up municipal electrical utilities to compete with older, larger, high-cost utilities.[3]
- Regulators often require the divestment of utility subunits of merging companies in order to preserve competition. To offset these divestments, many national power-marketing utilities have acquired utilities in other, non-contiguous, parts of the country.[4]
- As companies in the U.S. power industries merge, they are restructuring themselves from vertical to horizontal monopolies, and expanding through mergers to spread costs that cannot be cut in other ways.
- As utilities move from vertical to horizontal structures, they are choosing to specialize in their core strengths as they extend the scope of their markets.[5] Some utilities choose to become exclusively power transmission and retailing companies,[6] while other companies have chosen to market innovative energy convergence combinations such as gas and electricity, nationwide.[7]
- FERC is leading the state and federal agencies in renegotiating the refinancing of as much as $150 billion of obsolete, high-cost power production equipment, much of which represents the stranded cost of nuclear power plants.[8]
- The gas and electrical industries are rapidly merging into one.[9] This is occurring because U.S. utilities are adopting the improved technologies of energy conversion—the conversion of coal to electricity, gas to electricity, natural-gas to liquid natural gas, and gas to petroleum—so that they can sell the source of energy that has the lowest cost.[10] This implies a shift to gas- and coal-fired electrical generation equipment.[11]
- The coal, gas and electrical utilities industries mergers have created a new top tier of large national power marketing firms. These firms are willing to compete with the formerly monopolistic state utilities in most parts of the United States, selling whichever of the energy sources is the lowest cost.
- The innovations of utilities include efforts to enhance their ability to engage in national power marketing. Two of these innovations are national advertising campaigns focused on company image,[12] and product lines providing one-stop shopping.[13]
- U.S. power-marketing firms are investing abroad to help themselves compete in international markets where all forms of power are becoming commodities; where power is being traded by producers, users, and middle men; and where power is increasingly being securitized through contractual arrangements in such a way that market arbitrage can occur.[14]

These activities present examples of industrial activity where deregulation promises to compress the consolidating industry's shakeout period into a space of just a few years. The merger activity is huge in dollar terms because the power industries require large capital investments due to the nature of power production—construction of facilities is expensive—and also because the mergers are occurring rapidly. But waves of mergers and acquisitions in the United States are nothing new. Over the last ten years, the annual amount of mergers and acquisitions has become so large that they are now considered a normal part of corporate capital investment in the United States. Mergers and acquisitions have become so common that during the last fifteen years, they have averaged from 16 to 20 percent of the total internal capital investments of U.S. corporations.[15] Thomas Copeland and Fred Weston, report in their textbook that since beginning in the early 1970s, mergers also have been performing an important role in the allocation of resources for the entire U.S. economy. This is because divestitures or partial sales have represented as much as 40 to 50 percent of total merger activities.[16]

THE INTERNATIONALIZING FINANCIAL COMPONENT OF POWER MERGERS

What is unusual about the current wave of utilities mergers is its increasing international dimension. To see how this international component emerged, it is best first to review the background of the merger activity itself.

Following the recovery from the Reagan recession, there was a rapid release of pent-up merger potential in the United States which produced a continuing wave of mergers that included large numbers of utility combinations. The utilities mergers were, at first, a normal part of the greatest merger activity in the United States since 1985,[17] and then, later were a logical response to the U.S. deregulation of the U.S. power industry. The utilities found themselves merging to increase their share value and to develop markets and operations large and efficient enough to withstand the coming competition of deregulated power markets. Following the dictates of theory, the mergers that spilled abroad into foreign direct investments via acquisitions were mostly made for the first time, were mostly bid at very low rates, and in most cases, gave all appearances of being very profitable, outstripping the rate of returns available with U.S. merger partners.[18]

Markets Push Utilities Abroad

From a British and Australian point of view, what was unusual in these mergers was that over half of the merger activity came about solely through *U.S.* acquisitions of foreign plants. The speed and magnitude of this direct foreign investment and the fact that it appeared to have been determined primarily by the quest for an increased rate of return was new and surprising. The capital markets of the U.S. economy are probably the most competitive, efficient, and demanding in the world, and it was apparently their stringent return on investment requirements that pushed them abroad before it happened in other countries. This was true despite the fact that both Britain and Australia had deregulated their utilities before the United States. In fact, almost as soon as British and Australian utilities were privatized and deregulated, their new profit potential attracted U.S. bids. Such was the need of these U.S. firms that they acquired approximately 50 percent of the utility capacity in Britain and Australia from 1992 to 1997.

New Mergers Between Distant Utilities

Traditionally, mergers in the utilities industry were expected to be mergers between contiguous utilities. The objective would be to extend transmission networks and to produce the scale economies necessary for more efficient, better integrated, power production. But under FERC leadership, contiguous mergers are now being disapproved because they might cause an aggregation of market power that might prevent competition. Rather, current utilities mergers appear to have been to prompted by financial motives—the need to increase the utility's rate of return to the level expected by U.S. capital markets. Many of these mergers are with noncontiguous utilities or overseas utilities that offer an excellent rate of return through superior synergies. The nature of merger synergies, though, has changed as much as the nationalities of the merger partners, for the large mergers created the national power marketing firms that developed a simultaneous hunger for capital and new markets that the investment of that capital could guarantee. To keep the capital flowing, this top tier of firms needed a guarantee of a high rate of return.

To be competitive, the utilities needed lower costs based on newer, better plants, and in order to finance this new equipment, and to refinance and retire obsolete equipment, new capital investment was needed. To secure the capital necessary for this investment the utility needed a rate of return acceptable to the capital markets in which they issue stocks and bonds—approximately 15 percent. Most U.S. utilities were not earning this much. When favorable merger partners in the U.S. were exhausted, the utilities made offers to acquire private, deregulated, foreign power companies that had low costs, high growth, and favorable profit potential. The new production technologies of energy convergence, and the new market technologies of energy trading and arbitrage, meant that these mergers were not limited to contiguous utilities, domestic utilities, or utilities producing the same form of power.

Global Capital Markets
Breed Global Utilities

These conclusions suggest that many industries are discovering that size does not protect them from the demands of global capital markets. Firms of any nationality might be expected to respond to market financial dictates no matter where they are located. This observation suggests that further research is needed on the theory of foreign direct investments. Many of the the histories of foreign direct investment are rooted in theories of manufacturing corporations that became global in the 1960s and 1970s when cross-border operations were being discovered by large numbers of firms.[19] These theories and others that have followed may explain the initial surge of foreign direct investment, but now have become unnecessarily overcomplicated explanations of more recent foreign direct investment.[20]

Current foreign investment in the internationalizing utilities industry can be explained by more traditional theories of financial market capitalism. In the utilities industry, corporations appear to be striving for the rate of return that markets expect on their capital investment. Given the capital that these rates of return can attracts firms can invest in new, low-cost generating equipment. With low costs and enough capital for an advertising campaign, the firms can acquire new customers to purchase their power. Thus high rates of return and market capital provide the merging tier of national power-marketing firms with increasingly large market shares. These market shares are not yet disturbingly large though, because of the merger of state markets into national markets, and the merger of the markets for coal, gas, and electricity. These consolidations have dramatically increased the size of the market in which these large firms operation.

With these forces at work, the assumption that all large utilities are now operating under the scrutiny of global capital markets is a reasonable one, and the global utilities' drive for high return on investment satisfactorily explains most of their foreign direct investment and mergers and acquisitions activity in a very straightforward way. All that remains is an interesting perusal of the spectrum of new techniques that these global firms are using to cover the world.

The Globalization of British Power

As the world's power industry becomes global, Britain's power industry is becoming global too. The globalization of British power is occurring for many of the same reasons that spurred U.S. electrical utilities to acquire British partners—better profit opportunities abroad. Unlike the U.S.-European investment environment in the postwar period, British utilities are not counterattacking U.S. utilities in retaliation for their acquisition incursions in Britain.[21] Rather, British utilities are using the same strategies—the pursuit of better efficiency, better marketing, lower cost, and faster growing, deregulated markets—and they are scanning and investing globally.

Long Term Contracts Versus FDI

One of the powerful innovations built into the global spread of utilities is the use of long-term energy contracts as a substitute for foreign direct investment (FDI). Instead of building a factory to manufacture products for a foreign market, thereby substituting local production for trade goods, a utility might find it far more convenient to simply sell a foreign utility a twenty-five-year contract to deliver electricity.

Something like this was done by AES Corporation (AES) when it bought out GPU Corporation's (GPU) contracts with its high-cost independent suppliers of electricity destined for customers in New Jersey and Pennsylvania.[22] The investment proceeded in two stages: first AES purchased their market in the form of GPU's supply contracts, then they built a low-cost electrical generation plant to supply this market, but in neither case did they compete for their customers. Since AES wholesaled their electricity to GPU, GPU supplied AES with its new customers. In this deal, half of the investment was in the form of intangible twenty-five-year electricity supply contracts. Notice that a physical investment in the power plant was made only because this was the style of AES's operation. AES might just as easily have contracted the electricity from another low cost supplier and done the investment entirely through intangible energy contracts.

THE SECURITIZATION OF ENERGY

If you realize that long-term energy contracts can be the equivalent of a foreign direct investment and that a foreign direct investment in a large utility is often a merger or an acquisition, the lines between energy contracts, FDI, and mergers and acquisitions activities begin to blur. The merger that epitomizes such a situation is PacifiCorp's $5.8 billion bid for Energy Group PLC in June 1997.[23]

Pacificorp Swaps International Contracts

PacifiCorp was already a low-cost power marketer of coal and coal-fired electricity throughout the United States. The acquisition of Energy Group's coal and generating assets in Britain would have given PacifiCorp the ability to swap long-term coal and energy contracts not only within the United States, but also with overseas energy markets.[24] All that would be necessary to do that is that the trading utilities would need operations in the countries of the subsidiaries doing the trading. PacifiCorp would undoubtedly attempt to arbitrage these energy contracts, purchasing low and selling high whichever of the products produced the best profit. This ability to swap contracts internationally would also give PacifiCorp the ability to hedge risk in its energy markets. This is no trivial matter, for, because of PacifiCorp's size, each of its contracts might, if it wished, be the equivalent of a small foreign investment.

Oglethorpe Bets Half of Their Business

Another example of the power of long-term contracting in the new utilities industry is Oglethorpe Power's contract with Louisville Gas & Electric Company to provide Oglethorpe with over half of its electricity for the next fifteen years.[25] This contract means that Oglethorpe has wagered half the value of their firm on the electricity prices of the next fifteen years. Since their contract is the equivalent of a call on the price of electricity, if electricity prices do not rise above Oglethorpe's contract price—the equivalent of a call option's strike price—the consequences for them could be dire.

If firms like Oglethorpe were required to report the changes in value of a contract responsible for a supply of half their electrical power production, the capital market's perceived risk of their debt might go through the roof. In fact, the Financial Accounting Standards Board (FASB) had been considering a rule requiring that these changes in options values be reported as an expense on the income statement.[26] Perhaps with situations like that of Oglethorpe in mind, Fed chairman William Greenspan[27] and leading financial institutions[28] convinced the FASB to back off the rule change.

Although the FASB backed away from a financial reporting rule that would put the options risk of utilities out in the open, the risk did not go away. Notice that for a very large company like Louisville Gas & Electric, the company that sold the large electricity contract to Oglethorpe, it is much easier to hedge such contract risk. LG&E would simply hedge by matching up purchases and sales within certain price ranges. But for a smaller company like Oglethorpe, such hedging is not now readily available, or if it is it is not at all certain that Oglethorpe could hedge the contract and also maintain their rate of return. Pondering Oglethorpe's situation makes much clearer why a firm like Oglethorpe might want to contract the energy management skills of a firm like Enron, which is backed up by a $200 million dollar energy arbitraging computer system.

From Mergers and Acquisitions to Commodity Trading

It is true that corporations make bets on products, prices, and factory investments every day, but in the case of large utilities, the deals are so simple and the figures so large that the increasing securitization of energy makes their dealing in long-term international energy contracts appear to be a super form of commodities trading. As yet there is no secondary market for these contracts, so they are a lot like Eurobonds: if you buy an energy contract you must be prepared to hold it to maturity. But the pace of change in the internationalizing utilities is rapid and this creates the possibility that long-term energy contracts may become standardized, allowing a secondary market to develop.

Eurocommodities Trading
in Energy Contracts

Because of the large size and long term of most energy contracts, as well as the global spread of the utilities-energy industry, the contracts, like Eurobonds, would need a global market from the outset if the market were to achieve any reasonable degree of liquidity.[29] If a market in long-term energy contracts did develop, it would probably start out as an over-the-counter market in the London Eurosecurities market. If the contracts were traded in the London Euromarkets they would be denominated in various currencies depending on the origin of the energy or the nationality of the producer with the U.S. dollar as the probable dominating currency.

Although long-term energy contracts are presently being traded by energy producers and utilities, it would be difficult for a secondary market to develop. Should such a market develop it would have the same problems that the secondary markets in Eurobonds do.[30] Market makers in long-term energy contracts would be expected to hold short positions when the market was rising and long positions when the market was falling, so that they might be forced to take large capital losses over the long swings of energy price cycles. Market makers would also find it necessary to trade with investors who had superior information, for example, the large utilities and oil companies. This would mean that market makers would be vulnerable to the insider trading problems that spring from such an asymmetry of energy information. To make a market, brokers might have to take the losing side of a deal with a large oil company with inside information. These problems mean that profits might be thin, and capital requirements high for market makers. One way to finance these costs would be through the use of large trading spreads, but large spreads would scare away customers, many of whom might be large utilities and energy producers already possessing the expertise to trade among themselves.

NOTES

1. Robert J. Samuelson, "The Joy of Deregulation; A Bit Here, a Bit There and Soon We're Talking Big Savings, Say, $50 Billion a Year," *Newsweek*, February 3, 1997, p. 39, reports on the studies of Robert Crandall (Brookings Institution) and Jerry Ellig (George Mason Univ.), 1997.

2. James P. Miller and Benjamin A. Holden, "Northern States Power and Wisconsin Energy Called Off Their Merger, Citing the Unexpected Regulatory Complications They've Encountered," *Wall Street Journal*, May 19, 1997, pp. A1, A4; and James P. Miller and Benjamin A. Holden, "Wisconsin Energy Northern States Pact Rejected," *Wall Street Journal*, May 15, 1997, p. A1.

3. "Nassau County: A Bad Deal for Lynbrook," *Newsday*, March 13, 1997, and Andrea S. Halbfinger, "Study in Contrast, Lynbrook Village Meeting has Music, Cute Kids and Conflict," *Lynbrook Local News*, March 6, 1997, pp. 1, 3, 26.

4. Ross Kerber, "Auction of 18 Power Plants is Igniting Utilities' Interest," *Wall Street Journal*, February 6, 1997, pp. A1, B4.

5. R. Charles Moyer, "The Future of Electricity," *Business Economics*, October 1996, 31(4): 13-18.

6. Kerber, "Auction of 18 Power Plants," pp. A1, B4.

7. Agis Salpukis, "Enron Utility Merger Approved in Move to Expand Nationwide," *New York Times*, February 27, 1997, pp. D1, D4; and Benjamin A. Holden, "UtiliCorp and Peco, Aided by AT&T, To Launch One-Stop Utility Service," *Wall Street Journal*, July 24, 1997, pp. A1, A3.

8. Barnaby J. Feder, "The Nuclear Power Puzzle; Who Will Pay for a Generation of Expensive Plants?" *New York Times*, January 3, 1997, pp. D1, D3.

9. Charles H. Studness, "Converging Markets: The First Real Electric/Gas Merger," *Public Utilities Fortnightly*, October 1, 1996, 134(18): 21-25.

10. Ann Monroe, "The Looming Fusion of Power and Energy," *Investment Dealer's Digest*, July 29, 1996, 62(31): 16-17; Ann Monroe, "Big Deal Advances Fusion of Gas/Power Industries," *Invesment Dealer's Digest*, October 7, 1996, 62(41): 22; and Paul Feine, "Making the Mix," *Energy Economist*, July 1996, pp. 8-10.

11. Sack and Chewning, "Global Electricity Strategy"; McWilliams, "Energy: Gas to Oil," 132; Pearl and Fritsch, "Deep Pockets," pp. A1, A8; and Gundi Royle, "Gas in Europe—The Door to Competition is Pried Open," *Monthly Perspectives*, May 1996, and Richard H. K. Vietor, *Contrived Competition*, New York: McGraw Hill.

12. Salpukis, "Enron Utility Merger," pp. D1, D4.

13. Holden, "UtiliCorp and Peco," pp. A1, A3.

14. Benjamin A. Holden, "PacifiCorp Pursues Energy-Swap Plans," *Wall Street Journal*, June 16, 1997, pp. A1, B4; and Peter Coy and Gary McWilliams, "Electricity: the Power Shift Ahead," *Business Week*, December 2, 1996, pp. 78-82.

15. Thomas E. Copeland, and J. Fred Weston, Mergers Restructuring, and Corporate Control: Theory, chapter 19 in *Financial Theory and Corporate Policy*, 3d ed. (Reading, Massachusetts: Addison Wesley Publishing Company), p. 682; based on data from W. T. Grimm & Co., *Mergerstat Review 1985*, Chicago, 1986.

16. *Ibid.*

17. *Ibid.* Copeland and Weston point out that in constant dollar terms, 1985 was the first recent year to surpass the mergers and acquisitions record established in 1968.

18. John Doukas, "The Effect of Corporate Multinationalism on Shareholders' Wealth: Evidence from International Acquisitiions," Journal of Finance, December 1988, 43(5): 1181-1175.

19. Edward B.Flowers, "Oligopolistic Reaction in European and Canadian Direct Investment in the United States," *Journal of International Business Studies*, Fall/ Winter 1976, 7: 43-55, and Frederick T. Knickerbocker, *Oligopolistic Reaction and Multinational Enterprise* (Boston: Division of Research, Graduate School of Business Administration, Harvard University, 1973).

20. John H. Dunning, "Reappraising the Eclectic Paradigm in an Age of Alliance Capitalism," *Journal of International Business Studies*, Third Quarter 1995, 6(3): 461-491; John H. Dunning, "The Eclectic Paradigm of International Production: A Restatement and Some Possible Extensions," *Journal of International Business Studies*, Spring 1988, 19(1): 1-31; and John Dunning, "Trade Location of Economic Activity and the MNE: A Search for an Eclectic Approach," in Bertil Ohlin, Par-Ove Hesselborn, and Per Magnus Wijkman, eds., *The International Allocation of Economic Activity* (New York: Holmse and Meier, 1977), pp. 395-418.

21. Knickerbocker, *Oligopolistic Reaction.*

22. Agis Salpukas, "Utility Deal Aims to Cut Cost of Power," *New York Times*, February 5, 1997, pp. D1, D5.

23. Holden, "PacifiCorp Pursues Energy-Swap," pp. A1, B4.

24. *Ibid.*

25. Agis Salpukas, "Utility Seeks Partner to Aid Power Needs," *New York Times*, November 20, 1996, pp. D1, D2.

26. Floyd Norris, "Greenspan Opposes Accounting Change on Derivatives," *New York Times*, August 7, 1997, pp. D1, D8.

27. *Ibid.*

28. Floyd Norris, "Market Place: Some Big Financial Guns Attack a Proposal on Derivatives by the Accounting Standards Board," *New York Times*, August 1, 1997, pp. D1, D6.

29. See Maximo V. Eng, Francis A. Lees, and Laurence J. Mauer, International Bond Market, chapter 7 in *Global Finance*, 1st ed. (New York: Harper Collins College Publishers, 1995), pp. 181-220.

30. *Ibid.*, "International Bond Market, Secondary Trading," chapter 2, pp. 202-205.

Selected Bibliography

"Accounting Board to Adopt Derivative Rules," *New York Times*, August 12, 1997, pp. D1, D2.

Allanna, Sullivan. "Enron Deal Signals Trend in Utilities," *Wall Street Journal*, July 23, 1996, p. A3.

Bagli, Charles V. "ADT and Tyco Plan to Merge in $5.4 Billion Stock Swap," *New York Times*, March 18, 1997, pp. D1, D21.

Bagli, Charles V. "$2 Billion Deal Expected for Utility, Kansas City Power Said to Agree to Takeover," *New York Times*, February 7, 1997, pp. D1, D5.

Bagli, Charles V. "Conditions are Right for a Takeover Frenzy," *New York Times*, January 2, 1997, C3.

Bagli, Charles V. "A New Breed of Power Broker; Bankers are Changing the Shape of the Utility Industry," *New York Times*, December 10, 1996, pp. D1, D9.

Bagli, Charles V. "PG&E Will Buy 18 Power Plants in New England," *New York Times*, August 7, 1997, pp. D1, D2.

Bailey, Jeff. "Niagara Mohawk Plan is a Small Step Toward Easing Utilities' Power Woes," *Wall Street Journal*, March 12, 1997, pp. A1, A4.

Betts, Paul. "Enel, Enron Set to Form Joint Venture," *Financial Times*, June 4, 1997, p. 15.

Bloomberg News. "ARCO to Develop Gas Field in Indonesia," *New York Times*, September 3, 1997, pp. D1, D7.

Bloomberg News. "Enron to Pay $440 Million to settle Gas Dispute," *New York Times*, June 3, 1997, pp. D1, D6.

Boulton, Leyla. "Regulator Slams 'Huge' Water Dividend Rises," *Financial Times*, June 4, 1997, p. 8.

"Britain to Review Utility Acquisitions," *New York Times*, November 24, 1996, p. D12.

"Britain's Electricity Shocker," *Economist*, April 13, 1996, (339)7961: 14.

"British Utility Lifts Payout in Face of Bid," *New York Times*, December 11, 1996, p. D6.

Buckley, Christine. "PowerGen Spends £421M on Stakes in Two Far East Plants," *Financial Times*, May 23, 1997, pp. 26, 28.

"Calenergy Bid for Utility is Victorious; British Distributor is Latest Acquisi-tion," *New York Times*, December 25, 1996, pp. D1, D3.

Canedy, Dana. "Suitor Drops $1.9 Billion Bid for Utility," *New York Times*, Au-gust 16, 1997, pp. 35-36

Canedy, Dana. "Suitor Drops $1.9 Billion Bid for Utility," *New York Times*, Au-gust 16, 1997, pp. 35-36.

Cooper, Elizabeth. "Powerful Questions, BOT, LILCO Answer Questions, Com-ments at Public Hearing," *Lynbrook Local News*, February 6, 1997, pp. 1, 27.

Copeland, Thomas E. and Weston, J. Fred. Chapter 19, Mergers, Restructuring, and Corporate Control: Theory In *Financial Theory and Corporate Policy*, 3d ed. Reading: Addison-Wesley Publishing Company, 1988, pp. 680-682.

Coy, Peter and McWilliams, Gary. "Electricity: The Power Shift Ahead," *Busi-ness Week*, December 2, 1996, pp. 78-82.

DePalma, Anthony. "Ontario to Shut 7 Border Nuclear Reactors," *New York Times*, August 14, 1997, D1, p. A6.

"Dominion Agrees to Buy Power Utility," *New York Times*, November 14, 1996, pp. D1, D4.

Dornbusch, Rudi. "Economics Viewpoint: The Asian Juggernaut Isn't Really Slowing Down Dynamism," *Business Week*, July 14, 1997, p. 16.

Doukas, John. "The Effect of Corporate Multinationalism on Shareholders' Wealth: Evidence from International Acquisitions," *Journal of Finance*, De-cember 1988, 43(5): 1181-1175.

Doukas, John. "The Effect of Corporate Multinationalism on Shareholders' Wealth: Evidence from International Acquisitions," *Journal of Finance*, Vol. 43, No. 5 (December 1988), 1181-1175.

Doukas, John. "The Effect of Corporate Multinationalism on Shareholders' Wealth: Evidence from International Acquisitiions," *Journal of Finance*, Vol. 43, No. 5 (December 1988), 1181-1175.

Dunning, John H. "Reappraising the Eclectic Paradigm in an Age of Alliance Capitalism," *Journal of International Business Studies*, Third Quarter, 1995, 6(3): pp. 461-491.

Dunning, John H. "The Eclectic Paradigm of International Production: A Re-statement and Some Possible Extensions," *Journal of International Business Studies*, Vol. 19, No. 1 (Spring 1988), pp. 1-31.

Dunning, John. "Trade Location of Economic Activity and the MNE: A Search for an Eclectic Approach" In Bertil Ohlin, Par-Ove Hesselborn, and Per Magnus Wijkman, eds., *The International Allocation of Economic Activity*. New York: Holmse and Meier, 1977, pp. 395-418.

Eng, Maximo V.; Lees, Francis A.; and Mauer, Laurence J. chapter 7 "International Bond Market" In *Global Finance*, New York: Harper Collins College Publishers, 1995.

"Entergy Will Pay $2.1 Billion for a Utility, London Electricity," *New York Times*, December 19, 1996, pp. D1, D5.

Ewing, Terzah. "Enron and Portland General Reduce Stock-Swap Ration and Boost Rate Cuts," *Wall Street Journal*, April 15, 1997, p. A6.

Fairclough, Gordon. "Niagara Mohawk to Pay 19 Producers $4 Billion to Alter, End Energy Pacts," *Wall Street Journal*, March 11, 1997, pp. A1, A3.

Feder, Barnaby J. "The Nuclear Power Puzzle: Who Will Pay for a Generation of Expensive Plants?" *New York Times*, January 3, 1997, pp. D1, D3.

Feine, Paul. "Making the Mix," *Energy Economist*, July 1996, pp. 8-10.

Fialka, John J. "Using Savvy Tactics, Bristol, Va., Unplugs from a Federal Utili-ty," *New York Times*, May 27, 1997, pp. A1-A9.

Fialka, John J. "U.S. Plans to Sell Uranium-Enrichment Operations," *Wall Street Journal*, August 14, 1997, pp. A1, A14.

Flowers, Edward B. "Oligopolistic Reaction in European and Canadian Direct Investment in the United States," *Journal of International Business Studies*, Fall/Winter, 1976, 7: 43-55.

Fritsch, Peter. "Enron to Unveil Energy Alliance Inolving 11 Cities in Califor-nia," *Wall Street Journal*, January 15, 1997, pp. A1, B4.

Fritsch, Peter and Kline, Maureen. "Enron, Italian Utility ENEL Expected to An-nounce Power-Marketing Venture," *Wall Street Journal*, June 3, 1997, pp. A1, A3.

Gasparino, Charles. "Electric Giants File for Bankruptcy Protection," *New York Times*, May 19, 1997, pp. A1, A3.

Halbfinger, Andrea S. "Study in Contrast, Lynbrook Village Meeting has Music, Cute Kids and Conflict," *Lynbrook Local News*, March 6, 1997, pp. 1, 3, 26.

Hernandez, Raymond. "Pataki is Willing to Give Up Parts of Welfare Plan," *New York Times*, July 3, 1997, pp. A1, B4.

Holberton, Simon. "Deregulation Has Set U.S. Power Companies Scouting the World for Opportunities," *Financial Times*, February 25, 1997, p. 23.

Holberton, Simon. "Generators Plug in Abroad," *Financial Times*, May 26, 1997, p. 17.

Holden, Benjamin A. "Enron Corp. Has Accord to Buy Portland General," *Wall Street Journal*, July 22, 1996, p. A3.

Holden, Benjamin A. "LG&E to Buy KU for $1.43 Billion in Stock," *Wall Street Journal*, May 22, 1997, p. A6.

Holden, Benjamin A. "PacifiCorp is Likely to Finance Purchase in U.K. with Sale of Telecom Business," *Wall Street Journal*, June 17, 1997, p. B4.

Holden, Benjamin A. "PacifiCorp to Take $65 Million Charge on Currency Deals," *Wall Street Journal*, August 13, 1997, pp. A1, A6.

Holden, Benjamin A. "PacifiCorp Pursues Energy-Swap Plans," *Wall Street Journal*, June 16, 1997, pp. A1, B4.

Holden, Benjamin A. "PG&E Agrees to Buy Unit from Valero," *Wall Street Journal*, February 3, 1997, pp. A1, A3

Holden, Benjamin A. "UtiliCorp and Peco, Aided by AT&T, to Launch One-Stop Utility Service," *Wall Street Journal*, July 24, 1997, pp. A1, A3.

Holden, Benjamin. "Utility Chief Tests Strategy for New Era," *Wall Street Jour-nal*, May 29, 1996, p. B1.

Holden, Benjamin A. and Cooper, Helene. "Britain to Conduct Antitrust Review of PacifiCorp Deal," *Wall Street Journal*, August 4, 1997, pp. A1, B5.

"Home-Services Alliance Forms," *New York Times*, June 25, 1997, pp. D1, D5.

Ibrahim, Youssef M. "Blair Gains Tax Cut for Business, But the Rest of Britain Must Wait," *New York Times*, July 3, 1997, pp. A1, A5.

Jackson, Susan. "The Millstones Around NU's Neck," *Business Week*, December 2, 1996, p. 80.

Jenkins, Simon. "Monopoly Game Over," *Times*, May 21, 1997, p. 24.

Johnson, Kirk A. "Revived Long Island Finds Life after Military Contracts," *New York Times*, July 7, 1997, pp. A1, B6.

"Kansas City Power & Light Co.," *Wall Street Journal*, March 16, 1996, p. B6.

Kaserman, David L. "The Measurement of Vertical Economics and the Efficient Structure of the Electric Utility Industry," *Journal of Industrial Economics*, September 1991, 39(5): 483-502.

Kerber, Ross. "Auction of 18 Power Plants is Igniting Utilities' Interest," *Wall Street Journal*, February 6, 1997, pp. A1, B4.

Kerber, Ross. "Maine Yankee Power Plant's Owners Say Facility May Be Shut Permanently," *Wall Street Journal*, May 28, 1997, p. B4.

Kingstad, Jon Erik. "Merger Menace: Holding Companies and Overcapitaliza-tion," *Public Utilities Fortnightly*, January 15, 1996, pp. 42-45.

Knickerbocker, Frederick T. *Oligopolistic Reaction and Multinational Enter-prise*. Boston: Division of Research, Graduate School of Business Adminis-tration, Harvard University, 1973.

Kripalani, Manjeet. "India: 'You Have to Be Pushy And Aggressive," *Business Week*, September 24, 1997, p. 56.

Lakhman, Marina. "A Virtual Cleanup of Chernobyl, Israeli Software Markets Simulation Using Robots," *New York Times*, September 1, 1997, p. D3.

Lambert, Bruce. "Hot Issue in Lilco Takeover Talks: Who Pays for Shoreham?," *New York Times*, March 16, 1996, pp. 41-42.

Lambert, Bruce. "Stockholders of Lilco and Brooklyn Union Ratify Merger of Companies," *New York Times*, August 8, 1997, p. B15.

Lambert, Bruce. "Utility Sued Over Millstone 3 Closing," *New York Times*, Au-gust 8, 1997, B15.

Lambert, Bruce. "Warm Reception Turns Cold for Pataki's Plan to Take Over Lilco and Cut Rates," *New York Times*, July 17, 1997, p. B4.

Lipin, Steven and Fritsch, Peter. "Duke Power Plans to Acquire PanEnergy in Stock Transaction of About $7.7 Billion," *Wall Street Journal*, November 25, 1996, pp. A1, A3.

MacDonald, Elizabeth. "FASB May Back Off from Its Threat to Limit or End 'Poolings of Interest'," *Wall Street Journal*, July 1, 1997, A1, A4.

MacDonald, Elizabeth. "Merger Accounting Method Under Fire," *Wall Street Journal*, April 15, 1997, p. A4.

Macdonald, Elizabeth and Frank, Stephen E. "FASB Rejects Fed Chairman's Re-quest to Soften Proposed Rule on Derivatives," *Wall Street Journal*, August 12, 1997, pp. A1, A2, A9.

Marcial, Gene. "Everybody's Talking Takeover," *Business Week*, June 16, 1997, pp. 102-106.

Mathews, Anna Wilde, and Friedland, Jonathan. "U.S. Railroads are Making Tracks Overseas," *New York Times*, May 21, 1997, p. 4.

McWilliams, Gary. "Enron's Pipeline into the Future," *Business Week*, Decem-ber 2, 1996, p. 82.

McWilliams, Gary. "Energy: Gas to Oil: A Gusher for the Millennium?" *Busi-ness Week*, May 19, 1997, p. 132.

"Merrill Lynch Takes Dual Role in Electricity Deal," *Financial Times*, February 25, 1997, p. 23.

Michaels, Robert J. "Electric Utilities Mergers: The Answer or the Question?" *Public Utilities Fortnightly*, January 1, 1996, pp. 20-23.

Miller, James. "CalEnergy Ends $1.92 Billion Bid," *Wall Street Journal*, August 18, 1997, pp. A1, A4.

Miller, James P. and Lipin, Steven. "CalEnergy Launches Another Hostile Bid," *Wall Street Journal*, July 16, 1997, pp. A1, A3, A4.

Miller, James P. and Holden, Benjamin A. "Northern States Power and Wiscon-sin Energy Called Off Their Merger, Citing the Unexpected Regulatory Com-plications They've Encountered," *Wall Street Journal*, May 19, 1997, pp. A1, A4.

Miller, James P. and Holden, Benjamin A. "Wisconsin Energy Northern States Pact Rejected," *Wall Street Journal*, May 15, 1997, p. A1.

Monroe, Ann. "Big Deals Advance Fusion of Gas/Power Industries," *Investment Dealer's Digest*, October 7, 1996, 62(41): 22.

Monroe, Ann. "The Looming Fusion of Power and Energy," *Investment Dealer's Digest*, July 29, 1996, 62(31): 16-17.

Moyer, R. Charles. "The Future of Electricity," *Business Economics*, October 1996, 31(4). 13-18.

Myerson, Allen R. "Enron Will Buy Oregon Utility in Deal Valued at $2.1 Bil-lion," *New York Times*, July 23, 1996, p. D1.

Myerson, Allen R. "Enron Wins Pact to Supply Power from Wind Turbines," *New York Times*, May 27, 1997, p. A5.

Myerson, Allen R. "LG&E Energy Agrees to Buy Rival Utility in Kentucky," *New York Times*, May 22, 1977, pp. D1, D21.

"Nassau County: A Bad Deal for Lynbrook," *Newsday*, March 13, 1997.

Norris, Floyd. "Greenspan Opposes Accounting Change on Derivatives," *New York Times*, August 7, 1997, pp. D1, D8.

Norris, Floyd. "Market Place: Some Big Financial Guns Attack a Proposal on Derivatives by the Accounting Standards Board," *New York Times*, August 1, 1997, pp. D1, D6.

Parker-Pope, Tara. "U.K. Utilities Generate Takeover Frenzy, Latest Bid, Pow-ergen's for Midlands, Faces Hurdles," *Wall Street Journal*, September 18, 1995, pp. A16, A14.

Passell, Peter. "a Sea Change in Policy by the Trustbusters," The *New York Times*, March 20, 1997, pp. D1, D2.

Pearl, Daniel and Fritsch, Peter. "Deep Pockets: Natural Gas Generates Enthusi-asm and Worry in Oil-Soaked Mideast," *Wall Street Journal*, August 11, 1997, pp. A1, A8.

Perez-Pena, Richard. "In Plan to Cut L.I. Utility Rates, Biggest Savings Come Early," *New York Times*, March 27, 1997.

Perez-Pena, Richard. "Is This the End of Lilco? Silver Has the Crucial Vote," *New York Times*, July 13, 1997, pp. D1, B1, B7.

Perez-Pena, Richard. "Lilco's Hard Journey, Road to a State Takeover Began with Debacle of the Shoreham Plant," *New York Times*, July 21, 1997, p. B4.

Perez-Pena, Richard. "Pact Reached to Break Up Con Edison," *New York Times*, March 14, 1997, pp. B1, B4.

Perez-Pena, Richard. "Rate Cut Questions," *New York Times*, March 16, 1996, p. 42,

Perez-Pena, Richard. "State Officials Approve Partial Takeover of Lilco," *New York Times*, July 27, 1997, pp. D1, B6.

"Plan for Cut in Power Rates," *New York Times*, May 20, 1997, pp. A1, D1.

"Powergen of Britain May Bid for Midlands," *New York Times*, September 16, 1995, pp. 18, 32.

Quinn, Matthew C. "British Power Producer Spurns Southern's Overture," *Atlan-ta Constitution*, April 19, 1996, p. C1.

Quinn, Matthew C. "British Utility Seeks Own Merger in an Effort to Foil Southern Company," *Atlanta Constitution*, April 13, 1996, p. B2.

Quinn, Matthew C. "Southern Company Drops Attempt to Buy Second British Utility," *Atlanta Constitution*, May 9, 1996, p. E7.

Quinn, Matthew C. "Southern Company Pleased with British Venture in Natural Gas, *Atlanta Constitution*, April 30, 1996, p. D2.

Rabinovitz, Jonathan. "Hartford Says Utility Hid Nuclear Contamination," *New York Times*, September 16, 1997, pp. B1, B6.

Reed, Stanley. "Britain: Business Starts to Get the Message," *Business Week*, June 2, 1997, p. 58.

Reinhart, Andy. "The Business Week Global 1000: The Russians are Here the Russians are Here," *Business Week*, July 7, 1997, pp. 94-96.

Reuters. "Russian Utility to Try to Raise at Least $2 Billion," *New York Times*, May 17, 1997, pp. D1, D4.

Roll, Richard. "The Hubris Hypothesis of Corporate Takeovers," *Journal of Business*, 59 (April, 1986), pp. 197-216.

Rose, Matthew. "U.K. Utility Raises Bid for Rival in Latest Rebuff to a U.S. Suitor," *Wall Street Journal*, April 23, 1996, p. A19.

Royle, Gundi. "Gas in Europe—The Door to Competition is Pried Open," *Monthly Perspectives*, May 1996.

Sack, Judith B.and Chewning, Robert L. "Global Electricity Strategy: My Two Cents Worth (Or The Sustainable Price of Power)," Morgan Stanley, Interna-tional Investment Research, February 6, 1997.

Salpukas, Agis. "Big U.S. Utility Spreads Its Reach to Berlin," *New York Times*, May 24, 1997, p. D2.

Salpukas, Agis. "$1.9 Billion Hostile Bid for Utility," *New York Times*, July 16, 1997, pp. D1, D18.

Salpukas, Agis. "A 7.7 Billion Union of Gas, Electricity, Duke Power Gaining Panenergy's Sales Skill," *New York Times*, November 26, 1996, pp. D1, D6.

Salpukas, Agis. "Calenergy Offers to Buy British Utility," *New York Times*, Oc-tober, 19, 1996, pp. D1, D7.

Salpukas, Agis. "Enron Names Trading Chief as President of Company, Ap-pointment Reflects the Changing Market," *New York Times*, December 11, 1996, pp. D1, D6.

Salpukis, Agis. "Enron Utility Merger Approved in Move to Expand Nation-wide," *New York Times*, February 27, 1997, pp. D1, D4.

Salpukas, Agis. "Entergy Agrees to Manage Troubled Maine Nuclear Plant," *New York Times*, January 9, 1997, pp. D1, D6.

Salpukas, Agis. "Growing Natural-Gas Seller to Expand Electric Business; NGD Buying Destec, a Plant Operator," *New York Times*, February 19, 1997, pp. D1, D2.

Salpukas, Agis. "New Choices for Natural Gas; Retailers Find Users Puzzled as Industry Deregulates," *New York Times*, October 23, 1996, pp. D1, D4.

Salpukas, Agis. "Niagara Deal with Independents Could Reduce Price of Electri-city," *New York Times*, July 11, 1997, pp. D1, B5.

Salpukas, Agis. "Niagara Deal with Independents Could Reduce Price of Electri-city," *New York Times*, July 11, 1997, pp. D1, B5.

Salpukas, Agis. "Northeast Utilities Sues to Block Move by New Hampshire," *New York Times*, March 4, 1997, p. D8.

Salpukas, Agis. "Now Comes Hard Part in Utility Deal," *New York Times*, Feb-ruary 10, 1997, p. D2.

Salpukas, Agis. "Pacific Gas and Electric to Sell 4 Power Plants," *New York Times*, October 23, 1996, p. D4.

Salpukas, Agis. "Pacificorp is Said to Reach Deal to Buy British Utility," *New York Times*, June 12, 1997, pp. D1, D7.

Salpukas, Agis. "Says Pacificorp Considers a Bid," *New York Times*, June 11, 1997, pp. D1, D2.

Salpukas, Agis " U S Agency Moves to Ease Utility Mergers, with Deregula-tion and Price Competition, an Emphasis on Speed," *New York Times*, De-cember 19, 1996, pp. D1, D5.

Salpukas, Agis. "Utility Deal Aims to Cut Cost of Power," *New York Times*, February 5, 1997, pp. D1, D5.

Salpukas, Agis. "Utility Seeks Partner to Aid Power Needs," *New York Times*, November 20, 1996, pp. D1, D2.

Salpukas, Agis. "Utility Seeks to End Costly Pacts with Power Suppliers," *New York Times*, March 11, 1997, pp. D1, B8.

Salpukas, Agis. "When Electricity Goes Private, Deregulation May Change New York Power Authority," *New York Times*, July 11, 1997, pp. D1, D2, D3.

Samuelson, Robert J. "The Joy of Deregulation. a Bit Here, a Bit There and Soon We're Talking Big Savings. Say, $50 Billion a Year," *Newsweek*, Feb-ruary 3, 1997, 39, reports on the studies of Robert Crandall (Brookings Insti-tution) and Jerry Ellig George Mason Univ.), 1997.

Sayers, Michael. "The Death of Protected Profits," Morgan Stanley, Internation-al Investment Research, September 26, 1996.

"Southern Company: Plan to Purchase Utility Dropped After U.K. Action," *Wall Street Journal*, May 9, 1996, p. B4.

"Southern Company Unit Getting Role in Asia," *New York Times*, October 10, 1996, p. D4.

Specter, Michael. "Jeers Sting Mir Mission Control, Which Bemoans a Money Pinch," *New York Times*, August 20, 1997, p. 1.

Specter, Michael. "Russian Astronauts Insist Errors, While Human, Were on Earth," *New York Times*, August 17, 1997, pp. 1, 8.

Steinmetz, Greg. "Southern Co. Investment in Utility in Berlin May Be Unveiled Today," *Wall Street Journal*, May 13, 1997, pp. A1, A17.

Stout, David. "U.S. Accepts Suffolk's Bid to Import Electricity," *New York Times*, January 3, 1997, p. B2.

Strom, Stephanie. "International Business: British Reject 2 Power-Industry Takeovers; Utility Stocks, Up on Merger Speculation, Fall Sharply in Lon-don," *New York Times*, April 25, 1996, p. D7.

Studness, Charles H. "Converging Markets: The First Real Electric/Gas Mer-ger," *Public Utilities Fortnightly*, October 1, 1996, 134(18): 21-25.

Tagliabue, John. "Europe's Rail Freight on New Path: Revamping with Technol-ogy, Competition and U.S. Aid," *New York Times*, February 20, 1997, pp. D1, D15.

Taormina, Jo-Anne. "Letter from Lynbrook Village Community Relations: Vote on a Proposal to Allow the Village to Condemn a Portion of Lilco's Trans-mission Wires."

"Tax Row Mars Labor's Honeymoon," *Times*, May 18, 1997, p. 5.

Tejada, Carlos. "NGC to Acquire Destec for $127 Billion; Natural-Gas Concern Aims to Stake Out Position in Electicity Market," *Wall Street Journal*, Febru-ary 19, 1997, pp. A1, A2.

Truell, Peter. "in Financial Arena, Merger is the Game Major Players Seek Lift," *New York Times*, July 5, 1997, pp. A35, A38.

"Two Kansas City Utilities Agree to a Merger," *Los Angeles Times*, January 23, 1996, p. D2.

Van Horne, James C. *Financial Management and Policy*, 11th ed. Upper Saddle River, N.J.: Prentice Hall, 1997.

Wald, Matthew L. "An Industry Relishes Its Day in the Sun," *New York Times*, August 16, 1997, pp. 35-36.

Wald, Matthew L. "U.S.. to Put a Civilian Reactor to Military Use," *New York Times*, August 5,1997, pp. A1, A20.

Webster, Philip and Buckley, Christine. "Windfall Tax Faces Legal Challenge," *Times*, May 16, 1997, p. 1.

Wheatley, Jonathan. "U.S. Utilities Take Cemig Stake," *Financial Times*, May 29, 1997, p. 6.

Wighton, David. "Utilities Call for Equity in Windfall Tax," *Times*, May 20, 1997, p. 12.

Wilke, John R. "FTC Says Staples' Bid for Office Depot Sought to Remove Most Aggressive Rival," *Wall Street Journal*, May 20, 1997, pp. A1, C21.

Wilke, John R. "New Antitrust Rules May Ease Path to Mergers," *Wall Street Journal*, April 9, 1997, pp. A1, A3.

Wilke, John R. and Gruley, Bryan. "Merger Monitors: Acquisitions Can Mean Long-Lasting Scrutiny by Antitrust Agencies," *Wall Street Journal*, March 4, 1997, pp. A1, A10.

Yang, Catherine. "Regulators: This Top Trustbuster May Tone It Down," *Busi-ness Week*, October 7, 1996, p. 36.

Index

About the Author

EDWARD B. FLOWERS is Associate Professor in the Department of Economics and Finance at St. John's University's College of Business Administration. He is an attorney and former foreign affairs officer with the U.S. Treasury Department.

ISBN 1-56720-163-6

EAN

9 781567 201635

HARDCOVER BAR CODE